Lecture Notes in Mathematics

Edited by A. Dold and B. Eckmann

791

Karl Wilhelm Bauer
Stephan Ruscheweyh

Differential Operators for Partial Differential Equations and Function Theoretic Applications

Springer-Verlag
Berlin Heidelberg New York 1980

Authors
Karl Wilhelm Bauer
Institut für Mathematik
Technische Universität Graz
8010 Graz
Austria

Stephan Ruscheweyh
Mathematisches Institut
Universität Würzburg
8700 Würzburg
Federal Republic of Germany

AMS Subject Classifications (1980): 30C45, 30C60, 30C75, 30C80, 30D45, 30D50, 30D55, 30F35, 30G20, 30G30, 33A45, 35A20, 35C05, 35C10, 35C15, 35F05, 35J70, 35K05, 35Q05, 40C15, 40G10

ISBN 3-540-09975-1 Springer-Verlag Berlin Heidelberg New York
ISBN 0-387-09975-1 Springer-Verlag New York Heidelberg Berlin

Printed in Germany

Printing and binding: Beltz Offsetdruck, Hemsbach/Bergstr.
2141/3140-543210

Math

s-ep-

TABLE OF CONTENTS

PART I

Karl Wilhelm Bauer

Differential Operators for Partial Differential Equations

INTRODUCTION

In [40] G. Darboux introduced differential operators in connection with the Euler equation. In recent years this method for the representation of solutions of partial differential equations has become the object of increasing interest. Particularly this is based on the fact that these representations permit a detailed investigation of the function theoretic properties of the solutions. Especially in case of the differential equation

$$(*) \qquad (1+\varepsilon z\bar{z})^2 w_{z\bar{z}} + \varepsilon n(n+1)w = 0, \quad \varepsilon = \pm 1, \quad n \in \mathbb{N},$$

it was possible to generalize a number of statements of the classical function theory. In the first place by reason of the results proved by St. Ruscheweyh a function theory associated with the differential equation (*) could be developed. These results are treated in the subsequent contribution of St. Ruscheweyh. On the other hand the assertions obtained by differential operators permit a number of applications. Moreover, in various papers certain connections between differential and integral operators were investigated. However, a general characterization of those partial differential equations which permit representations of solutions by differential operators could not be found up to now. So much the more in this stage of the investigations a survey of the known results is of particular interest.
In the first chapter in the case of the differential equation

$$w_{z\bar{z}} + A(z,\bar{z})w_{\bar{z}} + B(z,\bar{z})w = 0$$

general conditions are derived for the appearance of solutions which may be represented by differential operators of order n operating on holomorphic respectively antiholomorphic functions. By certain additional conditions concerning the form of these differential operators one is led to the known representations of solutions. Subsequently various methods are characterized which permit to get corresponding representations of solutions for certain classes of other partial differential equations. Here, apart from solutions of the equation $h_{z\bar{z}} = 0$, also solutions of other elliptic or parabolic differential equations are used as generators.
The second chapter deals with several applications of the representa-

tion of solutions derived here. First a new representation of the spherical surface harmonics is treated. Besides, a corresponding class of functions is considered which arise in connection with the wave equation and may be called hyperboloid functions. Moreover, a representation of the surface harmonics of degree n in p dimensions is treated. The real and imaginary parts of certain classes of pseudo-analytic functions satisfy elliptic differential equations of the type considered here. Thus, it is possible to derive simple representations of these pseudo-analytic functions in simply connected domains and in the neighbourhood of isolated singularities. Moreover, these results permit various applications, for example, for a differential equation in the theory of functions of several complex variables, for a class of pseudo-analytic functions with a "sharp" maximum principle and for the determination of Vekua resolvents. By use of the results proved in Chapter I,5 in connection with the generalized Darboux equation it is possible to obtain representations for further classes of pseudo-analytic functions. By means of the complex potentials corresponding to these functions one is led to elliptic partial differential equations for which a representation of the solutions was not known up to now. Finally the integro-differential-operators treated in Chapter I,4 may be used for the representation of pseudo-analytic functions.

In Chapter II,4 we deal with a class of generalized Tricomi equations which lead to differential equations of the form considered here in the elliptic half-plane, whereas we get an Euler-Poisson-Darboux equation in the hyperbolic half-plane. Thus, in either case the solutions can be represented by differential operators. Moreover, by means of these representations it is possible to determine integral-free fundamental solutions in the large. These results are of particular interest since generalized Tricomi equations arise, for instance, in connection with the theory of transonic flow.

The assertions proved in Chapter I,2a may be used also for the representation of solutions of generalized Stokes-Beltrami systems. First, a system is treated which is closely related to a Stokes-Beltrami system which was considered by A. Weinstein in connection with the development of the generalized axially symmetric potential theory. Moreover, we deal with several systems of first-order partial differential equations to which we are led by certain functional-differential-relations for solutions of the Euler equation. Finally, by using the results of Chapter I,2a representations of solutions of the iterated equation of generalized axially symmetric potential theory are derived which arises in a number of physical problems.

Within each chapter, the sections, theorems, and formulas are numbered consecutively. If quoted, or referred to within the same chapter, only their own number is mentioned. Otherwise the number of the chapter is added, for instance, Theorem II,4 or (I,17).
I wish to thank Mrs. Heide Ditsios-Mack for her excellent typing of the manuscript.

August 1979 K. W. Bauer

CHAPTER I

Representation of solutions by differential operators

1) Polynomial operators for the differential equation $w_{z\bar{z}} + Aw_{\bar{z}} + Bw = 0$

a) Holomorphic generators

In the present paper we use the following notations. We set

$$z = x + iy$$

where x and y are real variables. i denotes the imaginary unit. Complex conjugates are denoted by bars, e.g.

$$\bar{z} = x - iy.$$

We use the formal differential operators

$$r = \frac{\partial}{\partial z} = (\quad)_z = \frac{1}{2}\left(\frac{\partial}{\partial x} - i\frac{\partial}{\partial y}\right) \tag{1}$$

and

$$s = \frac{\partial}{\partial \bar{z}} = (\quad)_{\bar{z}} = \frac{1}{2}\left(\frac{\partial}{\partial x} + i\frac{\partial}{\partial y}\right) . \tag{2}$$

Apart from the usual rules in case of differentiable functions we have

$$\bar{w}_z = \overline{(w_{\bar{z}})}, \qquad \bar{w}_{\bar{z}} = \overline{(w_z)} . \tag{3}$$

For a real-valued differentiable function, i.e. in case $w = \bar{w}$, we obtain

$$w_{\bar{z}} = \overline{(w_z)} . \tag{4}$$

Furthermore, for a holomorphic function g(z) we have

$$(5)\qquad \begin{cases} g_z = g', \quad g_{\bar{z}} = 0, \quad \bar{g}_{\bar{z}} = \overline{g'}, \\ (\operatorname{Re} g)_z = \frac{1}{2}g', \quad (\operatorname{Im} g)_z = -\frac{i}{2}g', \end{cases}$$

Re g denotes the real part of g, Im g denotes the imaginary part of g. Moreover, by (1) and (2) we get

$$(6)\qquad 4w_{z\bar{z}} = \Delta w,$$

where

$$\Delta = \frac{\partial^2}{\partial x^2} + \frac{\partial^2}{\partial y^2}$$

is the Laplace operator.
We consider partial differential equations of the form

$$w_{z\bar{z}} + a(z,\bar{z})w_z + b(z,\bar{z})w_{\bar{z}} + c(z,\bar{z})w = 0,$$

where a,b,c are given analytic functions in some domain. By a suitable transformation we can eliminate one of the two first derivatives. Therefore, we proceed from the normal form

$$(7)\qquad w_{z\bar{z}} + A(z,\bar{z})w_{\bar{z}} + B(z,\bar{z})w = 0.$$

We denote by D a simply connected domain of the complex plane, and we suppose that $A(z,\bar{z})$ and $B(z,\bar{z})$ are analytic in D.

By a solution of (7) in D we mean a function defined in D which has continuous partial derivatives up to the order two and satisfies equation (7) in D. By a theorem of E. Picard (cf. e.g.[88], p.162) such a solution is analytic, in particular there exist derivatives of all order.

Now we consider polynomial operators of order n. We set

$$(8)\qquad H(D) = \{\, g(z) \mid g(z) \text{ holomorphic in } D \,\}$$

and get by

$$P_n(r) = \sum_{k=0}^{n} a_k^*(z,\bar{z})r^k \tag{9}$$

the most general linear partial differential operator of order n on H(D). In view of future simple representations of solutions in place of (9) we use the form

$$P_n(r) = \sum_{k=0}^{n} a_k(z,\bar{z})R_o^k , \tag{10}$$

where the coefficients $a_k(z,\bar{z})$ are twice continuously differentiable functions in D and

$$R_o = a(z)r , \tag{11}$$

where a(z) is a holomorphic nonvanishing function in D.

Now we ask for all differential equations (7) which have solutions of the form

$$w = P_n g, \quad g \in H(D). \tag{12}$$

If we substitute (12) into (7), first we find $a_n = a_n(z)$. On account of (11) we put $a_n \equiv 1$ and obtain

$$\begin{cases} rsa_k + A\, sa_k + Ba_k + \dfrac{sa_{k-1}}{a} = 0 \\ \text{for } k = 0,1, \ldots, n-1 \text{ with } a_{-1} \equiv 0 \\ sa_{n-1} = -aB . \end{cases} \tag{13}$$

In characterizing the coefficients A and B in (7), in general, we are led to non-linear partial differential equations. In the sequel we consider some examples.

For n = 1 the system (13) reduces to

$$\begin{cases} rsa_o + A\, sa_o + Ba_o = 0 \\ sa_o = -aB . \end{cases} \tag{14}$$

If $B \neq 0$ in D, we get

$$(15) \qquad (\log B)_{z\bar{z}} + B + A_{\bar{z}} = 0.$$

Without loss of generality we may use $a \equiv 1$ and obtain the solution

$$(16) \qquad w = g' + a_o g,$$

where $a_o = A + (\log B)_z$. If w is a solution of (7) which may be represented in the form (16), we have

$$g = -\frac{w_{\bar{z}}}{B},$$

i.e. for a given solution w of this kind the generator g(z) is uniquely detérmined.

If the coefficient A in (7) satisfies certain conditions, we can derive further assertions. If, for example,

$$A_{\bar{z}} = B[\alpha(z)\overline{\beta(z)} - 1],$$

where $\alpha(z)$ and $\beta(z)$ are holomorphic nonvanishing functions in D, it follows

$$(\log B)_{z\bar{z}} + \alpha\bar{\beta}B = 0,$$

and by

$$G = \alpha\bar{\beta}B$$

we get the differential equation

$$(17) \qquad (\log G)_{z\bar{z}} + G = 0.$$

Setting here

$$\log G = 2W,$$

we obtain the Liouville equation

$$(18) \qquad 2W_{z\bar{z}} = -e^{2W}.$$

This is the special case of a differential equation which was investigated by G. Warnecke [104]. We quote some results of that paper as far as they are of interest for the following research.

Theorem 1

a) Let D^* be a simply connected domain of the complex plane. Let G be a solution of the differential equation (17)

$$(\log G)_{z\bar{z}} + G = 0$$

in D^*. Let D be a simply connected domain compact in D^*. Then, we can represent G in D by

$$G = \frac{-2\varphi'(z)\overline{\psi'(z)}}{(\varphi(z)+\overline{\psi(z)})^2}, \tag{19}$$

where $\varphi(z)$ and $\psi(z)$ are suitable holomorphic or meromorphic functions which satisfy the following conditions:

$$\left\{\begin{array}{l} \text{(i) } \varphi(z) \text{ and } \psi(z) \text{ have only a finite number of poles in D of at most first order.} \\ \text{(ii) } \varphi(z) \text{ and } \psi(z) \text{ have no common poles in D.} \\ \text{(iii) } (\varphi+\bar{\psi})\varphi'\psi' \neq 0 \text{ in D.} \end{array}\right. \tag{20}$$

b) Conversely, (19) represents a solution of (17) in D for each pair of holomorphic or meromorphic functions $\varphi(z)$ and $\psi(z)$ which satisfy the conditions (20).

c) Every real-valued solution of (17) defined in D may be represented by

$$G = \frac{2\varepsilon f'(z)\overline{f'(z)}}{[1+\varepsilon f(z)\overline{f(z)}]^2}, \quad \varepsilon = \pm 1, \tag{21}$$

where f(z) is a suitable holomorphic or meromorphic function in D which satisfies the conditions:

$$\left\{\begin{array}{l} \text{(i) } f(z) \text{ has only a finite number of poles in D of at most first order.} \\ \text{(ii) } (1+\varepsilon f\bar{f})f' \neq 0 \text{ in D.} \end{array}\right. \tag{22}$$

d) Conversely, (21) represents a real-valued solution of (17) in D for each holomorphic or meromorphic function f(z) which satisfies the conditions (22).

On the supposition for D we get the following

Theorem 2

a) The differential equation (7)

$$w_{z\bar{z}} + Aw_{\bar{z}} + Bw = 0, \qquad B \neq 0 \text{ in } D,$$

has a solution of the form

$$w = g' + a_o g, \qquad g(z) \in H(D), \tag{23}$$

if, and only of, the coefficients A and B satisfy the relation

$$(\log B)_{z\bar{z}} + B + A_{\bar{z}} = 0. \tag{24}$$

Then, the coefficient a_o in (23) follows by

$$a_o = A + (\log B)_z.$$

b) For every given solution w of (7) in D which may be represented by (23) the generator g(z) is uniquely determined by

$$g(z) = -\frac{w_{\bar{z}}}{B}.$$

c) If $A_{\bar{z}} = B[\alpha(z)\overline{\beta(z)}-1]$, where $\alpha(z)$ and $\beta(z)$ are holomorphic nonvanishing functions in D, we obtain by (24) with $G = \alpha\bar{\beta}B$

$$(\log G)_{z\bar{z}} + G = 0.$$

If the domain D satisfies the supposition in Theorem 1, the coefficient B may be represented in the form

$$B = \frac{-2\varphi'(z)\overline{\psi'(z)}}{\alpha(z)\overline{\beta(z)}[\varphi(z)+\overline{\psi(z)}]^2},$$

where $\varphi(z)$ and $\psi(z)$ satisfy the conditions (20).

If we impose on the coefficients $a_k(z,\bar{z})$ in (10) certain conditions, we may expect that the system (13) can be solved for arbitrary $n \in \mathbb{N}$.[1] We suppose $B \neq 0$ in D and set

$$a_k = c_k \eta^{n-k}, \quad n \geq 2, \tag{25}$$

where

$$c_k \in \mathbb{C}, \quad c_o \neq 0, \quad c_n = 1, \quad \eta = \eta(z,\bar{z}) \neq 0 \text{ in D.}$$

Then, (13) takes the form

$$c_k\{(n-k)\eta^{n-k-1}\eta_{z\bar{z}} + (n-k)(n-k-1)\eta^{n-k-2}\eta_z\eta_{\bar{z}} + A(n-k)\eta^{n-k-1}\eta_{\bar{z}} + \tag{26}$$

$$+ B\eta^{n-k}\} + \frac{c_{k-1}}{a}(n+1-k)\eta^{n-k}\eta_{\bar{z}} = 0$$

$$\text{for } k = 0,1, \ldots, n-1 \text{ with } c_{n-1} = 0,$$

$$c_{n-1}\eta_{\bar{z}} = -aB\,. \tag{27}$$

First we get $c_1, c_2, \ldots, c_{n-1} \neq 0$. Now we consider (26) with $k = 0$ and $k = 1$ and obtain by (27)

$$-\frac{\eta_z}{\eta^2} = \frac{c}{a}, \tag{28}$$

where

$$c = \frac{1}{n-1}\left[\frac{c_{n-1}}{n} - \frac{nc_o}{c_1}\right].$$

In the case $\eta_z \equiv 0$, i.e.

$$\eta = \overline{\gamma(z)}, \quad \gamma(z) \in H(D),$$

---

[1] We denote by $\mathbb{N}$, $\mathbb{Z}$, $\mathbb{R}$, and $\mathbb{C}$ the set of natural, integer, real, and complex numbers respectively. Moreover, we use $\mathbb{N}_o = \mathbb{N} \cup \{0\}$.

the coefficients A and B take the form

$$A = \frac{c_{n-1}}{na}\bar{\gamma}, \qquad B = -\frac{c_{n-1}}{a}\overline{\gamma'} .$$

Setting

$$\beta(z) = \frac{c_{n-1}}{n}\gamma \text{ and } \frac{1}{a} = \alpha(z), \qquad \alpha(z) \neq 0 \text{ in } D,$$

it follows by (7)

$$(29) \qquad w_{z\bar{z}} + \alpha(z)\overline{\beta(z)}w_{\bar{z}} - n\alpha(z)\overline{\beta'(z)}w = 0.$$

On the other hand, if we substitute

$$(30) \qquad w = \sum_{k=0}^{n} c_k \overline{\beta(z)}^{n-k} R_o^k g \quad \text{with} \quad R_o = \frac{1}{\alpha(z)}r$$

into (29), we get a solution by $c_k = \binom{n}{k}$. For a solution w of (29) which may be represented by (30) we obtain

$$T^{\mu}w = \sum_{k=0}^{n-\mu}\binom{n}{k}\frac{(n-k)!}{(n-k-\mu)!}\bar{\beta}^{n-k-\mu}R_o^k g$$

with

$$T = \frac{1}{\overline{\beta'}}\frac{\partial}{\partial\bar{z}},$$

as can be proved by induction. By $\mu = n$ we get

$$g(z) = \frac{T^n w}{n!} .$$

Supposing $c \neq 0$ in (28), the coefficients A and B take the form

$$A = \frac{\varphi'}{n(\varphi+\bar{\psi})}\left[\frac{c_{n-1}}{c} + n(n+1)\right], \qquad B = \frac{c_{n-1}\varphi'\overline{\psi'}}{c(\varphi+\bar{\psi})^2}$$

with

$$a(z) = \frac{1}{\varphi'(z)}$$

and $\varphi(z)$, $\psi(z) \in H(D)$, $(\varphi+\bar{\psi})\varphi'\psi' \neq 0$ in D. If we set

$$\lambda = \frac{1}{n}\left[\frac{c_{n-1}}{c} + n(n+1)\right],$$

it follows that the differential equation (7) takes on the form

$$(31) \qquad w_{z\bar{z}} + \frac{\lambda\varphi'}{\varphi+\bar{\psi}} w_{\bar{z}} - n(n+1-\lambda)\frac{\varphi'\overline{\psi'}}{(\varphi+\bar{\psi})^2} w = 0, \qquad \lambda \in \mathbb{C}, \qquad n \in \mathbb{N}.$$

On the other hand, if we substitute

$$(32) \qquad w = \sum_{k=0}^{n} c_k \eta^{n-k} R^k g, \qquad c_n = 1, \qquad R = \frac{1}{\varphi'}\frac{\partial}{\partial z},$$

with

$$\eta = \frac{1}{\varphi+\bar{\psi}}$$

into (31), we get a solution with

$$c_k = \frac{(-1)^{n-k} n!(n+1-\lambda)_{n-k}}{k!(n-k)!}, \quad {}^{2)}$$

where, on account of $c_k \neq 0$, we suppose that

$$\lambda \neq n+1, n+2, \ldots, 2n.$$

For a solution w of (31) which may be represented by (32) we obtain by induction

$$Q^{\mu}w = \sum_{k=0}^{n-\mu} c_k(-1)^{\mu} \frac{(n-k)!}{(n-k-\mu)!} \frac{R^k g}{(\varphi+\bar{\psi})^{n-k-\mu}}$$

with

---

2) Here we use the common notation

$$(c)_n = c(c+1) \ldots (c+n-1), \qquad n \in \mathbb{N},$$

$$(c)_0 = 1.$$

$$Q = \frac{(\varphi+\overline{\psi})^2}{\overline{\psi'}} \frac{\partial}{\partial \overline{z}} .$$

Finally, by $\mu = n$ we get

$$g(z) = \frac{Q^n w}{n!(n+1-\lambda)_n} .$$

Theorem 3

a) The differential equation (7)

$$w_{z\overline{z}} + Aw_{\overline{z}} + Bw = 0, \quad B \neq 0 \text{ in } D,$$

has a solution of the form

$$w = \sum_{k=0}^{n} c_k \eta^{n-k} R_0^k g, \quad g(z) \in H(D), \quad n \geq 2, \tag{33}$$

in D with

$$R_0 = a(z)\frac{\partial}{\partial z}, \quad a(z) \in H(D), \quad a(z) \neq 0 \text{ in } D,$$

$$c_k \in \mathbb{C}, \quad c_0 \neq 0, \quad c_n = 1, \quad \eta(z,\overline{z}) \neq 0 \text{ in } D,$$

if, and only if,

1) $A = \alpha(z)\overline{\beta(z)}, \quad B = -n\alpha(z)\overline{\beta'(z)}$

with

(i) $\alpha(z), \; \beta(z) \in H(D)$,
(ii) $\alpha\beta' \neq 0$ in D,

or if

2) $A = \dfrac{\lambda\varphi'}{\varphi+\overline{\psi}}, \quad B = -n(n+1-\lambda)\dfrac{\varphi'\overline{\psi'}}{(\varphi+\overline{\psi})^2}$

with

(i) $\varphi(z), \; \psi(z) \in H(D)$,
(ii) $(\varphi+\overline{\psi})\varphi'\psi' \neq 0$ in D,
(iii) $\lambda \notin \{n+1, n+2, \ldots, 2n\}$.

b) The solutions (33) of the differential equation

$$(34) \qquad w_{z\bar{z}} + \alpha(z)\overline{\beta(z)}\, w_{\bar{z}} - n\alpha(z)\overline{\beta'(z)}w = 0, \qquad n \in \mathbb{N},$$

may be represented in the form

$$(35) \qquad w = \sum_{k=0}^{n} \binom{n}{k} \bar{\beta}^{n-k} R_o^k g, \qquad R_o = \frac{1}{\alpha(z)} \frac{\partial}{\partial z}, \qquad g \in H(D).$$

c) For every given solution w of (34) which can be represented by (35) the generator g(z) is uniquely determined by

$$(36) \qquad g(z) = \frac{T^n w}{n!}, \qquad T = \frac{1}{\overline{\beta'}} \frac{\partial}{\partial \bar{z}}.$$

d) The solutions (33) of the differential equation

$$(37) \qquad w_{z\bar{z}} + \frac{\lambda\varphi'}{\varphi+\bar{\varphi}} w_{\bar{z}} - n(n+1-\lambda) \frac{\varphi'\overline{\varphi'}}{(\varphi+\bar{\varphi})^2} w = 0$$

may be represented in the form

$$(38) \qquad w = \sum_{k=0}^{n} (-1)^{n-k} \binom{n}{k} (n+1-\lambda)_{n-k} \frac{R^k g}{(\varphi+\bar{\varphi})^{n-k}}, \qquad R = \frac{1}{\varphi'} \frac{\partial}{\partial z}.$$ [3)]

e) For every given solution of (37) which can be represented in the form (38) the generator g(z) is uniquely determined by

$$(39) \qquad g(z) = \frac{Q^n w}{n!(n+1-\lambda)_n}, \qquad Q = \frac{(\varphi+\bar{\varphi})^2}{\overline{\varphi'}} \frac{\partial}{\partial \bar{z}}.$$

[3)] Considering Bergman integral operators with polynomial kernels, E. Kreyszig [81] was led to the differential equation

$$w_{z_1 z_2} + \frac{m}{kz_1+\psi(z_2)} w_{z_1} - n[(n+1)k-m] \frac{\psi'(z_2)}{[kz_1+\psi(z_2)]^2} w = 0.$$

If we set $z_1 = z$ and $z_2 = \bar{z}$, we get a special case of (37) with $\lambda = \frac{m}{k}$ and $\varphi(z) = kz$.

b) Antiholomorphic generators

We consider the partial differential operator

$$\overline{P}_m(s) = \sum_{k=0}^{m} b_k(z,\overline{z})s^k, \qquad m \in \mathbb{N},$$

where the coefficients $b_k(z,\overline{z})$ are twice continuously differentiable functions in D. We ask for all differential equations (7) which have solutions of the form

$$w = \overline{P}_m\overline{f}$$

with $f(z) \in H(D)$. Substituting

$$w = b_1\overline{f'} + b_o\overline{f}$$

into (7), we get in the case $m = 1$

$$rb_1 + b_1A = 0, \tag{40}$$

$$rsb_o + Asb_o + b_oB = 0, \tag{41}$$

$$rsb_1 + Asb_1 + b_1B + rb_o + b_oA = 0. \tag{42}$$

Differentiating (40) with respect to $\overline{z}$ and substituting the result into (42), it follows

$$b_1(B-sA) + rb_o + b_oA = 0. \tag{43}$$

First, we consider the case

$$B - sA \equiv 0.$$

Then by (40) and (43) we obtain

$$b_1 = \overline{\alpha(z)}b_o, \qquad \alpha(z) \in H(D),$$

and $w = b_o\overline{f_1}$ with $f_1 = \alpha f' + f$.

If $B - sA \neq 0$ in D, it follows by (40) - (42)

$$[\log(B-A_{\bar{z}})]_{z\bar{z}} + B = 2A_{\bar{z}} ,$$

$$b_o = sb_1 + b_1[\log(B-A_{\bar{z}})]_{\bar{z}} ,$$

$$rb_1 + b_1 A = 0,$$

and

$$w = (b_1\bar{f})_{\bar{z}} + [\log(B-A_{\bar{z}})]_{\bar{z}}(b_1\bar{f}) .$$

For a given solution w of this kind the function $b_1\bar{f}$ is uniquely determined by

$$b_1\bar{f} = \frac{Aw+w_z}{A_{\bar{z}}-B} .$$

Also in the present case we can derive further assertions about the coefficients A and B if additional conditions appear. If, for example,

$$A_{\bar{z}} = \frac{\alpha(z)\overline{\beta(z)}-1}{2\alpha(z)\overline{\beta(z)} - 1} B$$

with

$$\alpha(z), \beta(z) \in H(D) \quad \text{and} \quad \alpha\beta(2\alpha\bar{\beta}-1) \neq 0 \text{ in } D,$$

we are again led to the differential equation (17)

$$(\log G)_{z\bar{z}} + G = 0$$

by

$$G = \frac{B}{2\alpha\bar{\beta}-1} .$$

Theorem 4

a) The differential equation (7)

$$w_{z\bar{z}} + Aw_{\bar{z}} + Bw = 0, \qquad B - A_{\bar{z}} \neq 0 \text{ in } D,$$

has a solution of the form

$$w = b_1\overline{f'} + b_0\overline{f}, \qquad f \in H(D), \tag{44}$$

if, and only if, the coefficients A and B satisfy the relation

$$[\log(B-A_{\bar{z}})]_{z\bar{z}} + B = 2A_{\bar{z}} . \tag{45}$$

Then, the coefficients b_1 and b_0 follow by

$$b_{1,z} + b_1 A = 0, \qquad b_0 = b_{1,\bar{z}} + b_1[\log(B-A_{\bar{z}})]_{\bar{z}} ,$$

and the representation (44) takes on the form

$$w = (b_1\overline{f})_{\bar{z}} + [\log(B-A_{\bar{z}})]_{\bar{z}}(b_1\overline{f}) . \tag{46}$$

b) For every given solution of (7) which may be represented by (44) the function $b_1\overline{f}$ is uniquely determined by

$$b_1\overline{f} = \frac{Aw + w_z}{A_{\bar{z}} - B} .$$

c) If

$$A_{\bar{z}} = \frac{\alpha(z)\overline{\beta(z)} - 1}{2\alpha(z)\overline{\beta(z)} - 1} B$$

with

$$\alpha(z), \ \beta(z) \in H(D) \text{ and } \alpha\beta(2\alpha\overline{\beta} - 1) \neq 0 \text{ in } D,$$

the conditions (45) yields

$$(\log G)_{z\bar{z}} + G = 0$$

with $G = B(2\alpha\bar{\beta} - 1)^{-1}$. If the domain D satisfies the supposition in Theorem 1, the coefficient B can be represented in the form

$$B = \frac{-2(2\alpha\bar{\beta}-1)\varphi'\overline{\psi'}}{(\varphi+\bar{\psi})^2},$$

where the functions $\varphi(z)$ and $\psi(z)$ satisfy the conditions (22).

d) In the case $B - A_{\bar{z}} \equiv 0$ the differential equation (7) has a solution of the form (44) if

$$b_{o,z} + b_o A = 0, \quad b_1 = \overline{\alpha(z)} b_o, \quad \alpha(z) \in H(D).$$

Then, the representation (44) reduces to $w = b_o\overline{f_1}$, $f_1(z) \in H(D)$.

Analogous to (25) for $m \geq 2$ we use the operator

$$\sum_{k=0}^{m} d_k \tau^{\mu-k} S_o^k$$

with

$$d_k \in \mathbb{C}, \quad d_o d_m \neq 0, \quad \mu \in \mathbb{Z}, \quad \tau = \tau(z,\bar{z}), \quad S_o = \overline{d(z)} s,$$

where $d(z)$ is a holomorphic nonvanishing function in D, whereas $\tau(z,\bar{z})$ represents a twice continuously differentiable function which does not vanish in D.

Now we ask for all differential equations (7) with $B \neq 0$ in D which have solutions of the form

$$w = \sum_{k=0}^{m} d_k \tau^{\mu-k} S_o^k \bar{f}, \quad f \in H(D). \tag{47}$$

Substituting (47) into (7), we get

$$(48) \qquad A = (m-\mu)\,\frac{r\tau}{\tau}\,,$$

$$(49) \qquad B = \frac{-\mu}{\tau^2}[(m-1)r\tau s\tau + \tau rs\tau],$$

$$(50) \qquad kd_k[r\tau s\tau(m+\mu-1-k)+\tau rs\tau] = d_{k-1}\,\frac{\tau^2 r\tau}{\bar{d}}(m+1-k) \text{ for } k = 1, \ldots, m,$$

and $d_1, \ldots, d_{m-1} \neq 0$, $r\tau \neq 0$ in D. For $k = 1$ and $k = 2$ we obtain

$$-\frac{s\tau}{\tau^2} = \frac{c}{\overline{d(z)}}\,, \qquad c = \frac{m-1}{2}\,\frac{d_1}{d_2} - m\,\frac{d_o}{d_1}\,.$$

Because of $B \neq 0$ the case $c = 0$ falls away. For $c \neq 0$ we set

$$d(z) = \frac{1}{\psi'(z)}\,.$$

It follows that the coefficients A and B have necessarily the form

$$A = \frac{(\mu-m)\varphi'}{\varphi + \bar{\psi}}\,, \qquad B = -\mu(m+1)\frac{\varphi'\overline{\psi'}}{(\varphi+\bar{\psi})^2}$$

with $\varphi(z), \psi(z) \in H(D)$ and $(\varphi+\bar{\psi})\varphi'\psi' \neq 0$ in D. On the other hand, if we substitute

$$w = \sum_{k=0}^{m} d_k\,\frac{S^k\bar{f}}{(\varphi+\bar{\psi})^{n-k}}\,, \qquad S = \frac{1}{\overline{\psi'}}\,\frac{\partial}{\partial\bar{z}}\,,$$

into the differential equation

$$(51) \qquad w_{z\bar{z}} + (\mu-m)\frac{\varphi'}{\varphi+\bar{\psi}}\,w_{\bar{z}} - \mu(m+1)\frac{\varphi'\overline{\psi'}}{(\varphi+\bar{\psi})^2}\,w = 0,$$

we get a solution with

$$d_k = (-1)^{m-k}\binom{m}{k}(\mu+1)_{m-k}\,, \qquad \mu \neq -1,-2, \ldots, -m.$$

For a given solution of (51) which can be represented in the above-mentioned form the function f(z) is uniquely determined. Applying the operator

$$P = \frac{(\varphi+\overline{\psi})^2}{\varphi'} \frac{\partial}{\partial z},$$

first we find

$$P^{\sigma}[w(\varphi+\overline{\psi})^{\mu-m}] = \sum_{k=0}^{m-\sigma} (-1)^{m-k+\sigma} \binom{m}{k} (\mu+1)_{m-k} \frac{(m-k)!}{(m-k-\sigma)!} \frac{S^k\overline{f}}{(\varphi+\overline{\psi})^{m-k-\sigma}}$$

and then by $\sigma = m$

$$f(z) = \frac{\overline{P^m[w(\varphi+\overline{\psi})^{\mu-m}]}}{m!(\mu+1)_m}.$$

Theorem 5

a) The differential equation (7)

$$w_{z\overline{z}} + Aw_{\overline{z}} + Bw = 0, \quad B \neq 0 \text{ in } D,$$

has a solution of the form

$$w = \sum_{k=0}^{m} d_k \tau^{\mu-k} S_o^k \overline{f}, \quad m \geq 2, \quad f \in H(D), \tag{52}$$

with

$$S_o = \overline{d(z)} \frac{\partial}{\partial \overline{z}}, \quad d(z) \in H(D), \quad d(z) \neq 0 \text{ in } D,$$

$$d_k \in \mathbb{C}, \quad d_o d_m \neq 0, \quad \mu \in \mathbb{Z}, \quad \tau(z,\overline{z}) \neq 0 \text{ in } D,$$

if, and only if,

$$A = \frac{(\mu-m)\varphi'}{\varphi+\overline{\psi}}, \quad B = -\mu(m+1)\frac{\varphi'\overline{\psi'}}{(\varphi+\overline{\psi})^2},$$

where

(i) $\varphi(z), \psi(z) \in H(D)$,

(ii) $(\varphi+\bar{\psi})\varphi'\psi' \neq 0$ in D,

(iii) $\mu \notin \{-1, -2, \ldots, -m\}$.

b) The solution (52) of the differential equation

$$(53) \qquad w_{z\bar{z}} + (\mu-m)\frac{\varphi'}{\varphi+\bar{\psi}} w_{\bar{z}} - \mu(m+1)\frac{\varphi'\overline{\psi'}}{(\varphi+\bar{\psi})^2} w = 0$$

can be represented in the form

$$(54) \qquad w = \sum_{k=0}^{m} (-1)^{m-k} \binom{m}{k} (\mu+1)_{m-k} \frac{S^k \bar{f}}{(\varphi+\bar{\psi})^{\mu-k}}, \quad S = \frac{1}{\overline{\psi'}} \frac{\partial}{\partial \bar{z}}, \ f(z) \in H(D).$$

c) For every given solution w of (53) which may be represented in the form (54) the function f(z) is uniquely determined by

$$(55) \qquad f(z) = \overline{\frac{P^m[w(\varphi+\bar{\psi})^{\mu-m}]}{m!(\mu+1)_m}}, \quad P = \frac{(\varphi+\bar{\psi})^2}{\varphi'} \frac{\partial}{\partial z} .$$

2) The differential equation $\omega^2 w_{z\bar{z}} + (n-m)\varphi'\omega w_{\bar{z}} - n(m+1)\varphi'\overline{\psi'}w = 0$

a) A general representation theorem for the solutions defined in simply connected domains

Considering the results in Theorem 3 and Theorem 5, we can determine all differential equations (7) which habe solutions of the form (38) as well as such of the form (54). In this case the quantities λ, μ, n, and m are to satisfy the relations

$$\lambda = \mu - m \quad \text{and} \quad n(n+1-\lambda) = \mu(m+1).$$

Consequently,

$$\mu = n, \quad \lambda = n - m$$

and the differential equation takes on the form

$$(56) \qquad w_{z\bar{z}} + (n-m)\frac{\varphi'}{\varphi+\overline{\psi}} w_{\bar{z}} - n(m+1)\frac{\varphi'\overline{\psi'}}{(\varphi+\overline{\psi})^2} w = 0 .$$

For the solutions of this differential equation further results are known (cf.[18]). Among other things it was possible to determine all solutions defined in simply connected domains. Moreover, further representations by differential operators were given, where also solutions of the differential equation $h_{z\bar{z}} = 0$ arise as generators. In the following theorem we summarize some of these results and refer the reader to [18] for further details.

Theorem 6

a) For every solution of the differential equation (56)

$$w_{z\bar{z}} + (n-m)\frac{\varphi'}{\varphi+\overline{\psi}} w_{\bar{z}} - n(m+1)\frac{\varphi'\overline{\psi'}}{(\varphi+\overline{\psi})^2} w = 0, \qquad n,m \in \mathbb{N}_0,$$

with

$$\varphi(z), \psi(z) \in H(D) \quad \text{and} \quad (\varphi+\overline{\psi})\varphi'\psi' \neq 0 \text{ in } D,$$

defined in D, there exist two functions $g(z)$, $f(z) \in H(D)$, such that

$$
w = D_n g + D_m^* \overline{f} \tag{57}
$$

$$
= \omega^{m+1} \left[R^n \left[\frac{m!\, g(z)}{n!\, \omega^{m+1}} \right] + S^m \left[\frac{n!\, \overline{f(z)}}{m!\, \omega^{n+1}} \right] \right]
$$

$$
= \omega^{m+1} S^m R^n \frac{h}{\omega}
$$

with

$$
h(z,\overline{z}) = \frac{(-1)^m}{n!} g(z) + \frac{(-1)^n}{m!} \overline{f(z)}, \qquad \omega = \varphi + \overline{\varphi} ,
$$

where D_n respectively D_m^* denote the differentialoperators

$$
D_n = \sum_{k=0}^{n} \frac{(-1)^{n-k}(n+m-k)!}{k!(n-k)!\,\omega^{n-k}} R^k , \tag{58}
$$

$$
D_m^* = \sum_{k=0}^{m} \frac{(-1)^{m-k}(n+m-k)!}{k!(m-k)!\,\omega^{n-k}} S^k . \tag{59}
$$

b) Conversely, for arbitrary functions $g(z)$, $f(z) \in H(D)$ (57) represents a solution of (56) in D.

c) For every given solution w of (56) in D the functions

$$
R^{n+m+1} g = \frac{n!\, P^{m+1}(\omega^{n-m} w)}{m!\, \omega^{n+m+2}} , \qquad P = \frac{\omega^2}{\varphi'} \frac{\partial}{\partial z} , \tag{60}
$$

and

$$
S^{n+m+1} \overline{f} = \frac{m!\, Q^{m+1} w}{n!\, \omega^{n+m+2}} , \qquad Q = \frac{\omega^2}{\overline{\varphi'}} \frac{\partial}{\partial \overline{z}} \tag{61}
$$

are uniquely determined.

d) For every given solution w the generators $g(z)$ and $f(z)$ are only determined up to a polynomial $p(\varphi)$ of degree n+m. We obtain the most general pair of generators $\tilde{g}(z)$ and $\tilde{f}(z)$ by

(62) $$\tilde{g}(z) = g(z) + p(\varphi),$$

(63) $$\tilde{f}(z) = f(z) + \frac{m!}{n!}(-1)^{n+m+1}\,\overline{p(-\bar{\psi})}\,.$$

e) For every solution w of (56) in D which can be represented by only one generator g(z) or f(z) this function is uniquely determined by

(64) $$g(z) = \frac{Q^n w}{(n+m)!},$$

(65) $$f(z) = \frac{\overline{P^m(\omega^{n-m} w)}}{(n+m)!}\,.$$

Corollary

Because of (60) and (61) for every solution w of (56) which is defined in a (not necessarily simply connected) domain D the functions

$$R^{n+m+1}g \quad \text{and} \quad \overline{S^{n+m+1}\bar{f}}$$

are uniquely determined in each point of D and represent globally unique holomorphic functions in D.

b) General expansion theorems for the solutions in the neighbourhood of isolated singularities

The results of Theorem 6 and the Corollary offer the possibility to deduce general expansion theorems for the solutions of (56) in the neighbourhood of isolated singularities. In the sequel the corresponding considerations are sketched briefly (for further details the reader is directed to [18]).

Let $\varphi(z)$ and $\psi(z)$ be holomorphic in the disk

$$U(z_o) = \{\, z \mid |z-z_o| < \rho \,\}\,.$$

Let w be a solution of (56) defined and unique in the punctured open disk

$$\dot{U}(z_o) = \{ z \mid 0 < |z-z_o| < \rho \} \quad .$$

Then, from the Corollary it follows for the generators

$$R^{n+m+1}g = \sum_{-\infty}^{\infty} a_k^*(z-z_o)^k, \quad a_k^* \in \mathbb{C},$$

$$\overline{S^{n+m+1}\overline{f}} = \sum_{-\infty}^{\infty} b_k^*(z-z_o)^k, \quad b_k^* \in \mathbb{C},$$

and by indefinite integration we get

$$g(z) = g_1(z) + p_1(\varphi)\log(z-z_o),$$

$$f(z) = f_1(z) + p_2(\psi)\log(z-z_o)$$

with

$$p_1(\varphi) = \sum_{\mu=0}^{n+m} c_\mu \varphi^\mu, \quad c_\mu \in \mathbb{C},$$

$$p_2(\psi) = \sum_{\mu=0}^{n+m} d_\mu \psi^\mu, \quad d_\mu \in \mathbb{C},$$

whereas $g_1(z)$ and $f_1(z)$ denote Laurent series about z_o. The polynomials $p_1(\varphi)$ and $p_2(\psi)$ have to satisfy the condition

$$D_n p_1 - D_m^* \overline{p_2} = 0$$

since $w = D_n g + D_m^* \overline{f}$ represents a unique solution in $\dot{U}(z_o)$.

$$w^* = D_n p_1 - D_m^* \overline{p_2} = D_n p_1 + D_m^*(-\overline{p_2})$$

is a solution of (56) defined in $U(z_o)$ which by Theorem 6,d can also be represented in the form

$$w^* = D_n \tilde{p}_1 + D_m^* \overline{\tilde{p}_2}$$

with

$$\tilde{p}_1(\varphi) = p_1(\varphi) + p(\varphi),$$

$$\tilde{p}_2(\psi) = -p_2(\psi) + \frac{m!}{n!}(-1)^{n+m+1}\overline{p(-\bar{\psi})} .$$

If we choose

$$p(\varphi) = -p_1(\varphi),$$

it follows

$$\tilde{p}_1(\varphi) \equiv 0$$

and

$$\tilde{p}_2(\psi) = -\sum_{\mu=0}^{n+m} [d_\mu + \frac{m!}{n!}(-1)^{n+m+1}\overline{c_\mu}]\,\psi^\mu . \tag{66}$$

With

$$w^* = D_m^*\overline{\tilde{p}_2} = 0$$

we get by Theorem 6,e

$$\tilde{p}_2 \equiv 0$$

and by (66)

$$p_2(\psi) = \frac{m!}{n!}(-1)^{n+m}\ \overline{p_1(-\bar{\psi})} .$$

<u>Theorem 7</u>

Let w be a solution of (56) in $\dot{U}(z_o)$ with an isolated singularity at z_o. Then, w can be represented in $U(z_o)$ by

$$w = D_n g + D_m^*\bar{f}$$

with the generators

$$g(z) = g_1(z) + q(\varphi)\log(z-z_o),$$

$$f(z) = f_1(z) + \frac{m!}{n!}(-1)^{n+m}\,\overline{q(-\bar{\psi})}\log(z-z_o),$$

where $g_1(z)$ and $f_1(z)$ are holomorphic and unique functions in $\dot{U}(z_o)$ whereas $q(\varphi)$ represents an arbitrary polynomial in φ of degree $n+m$.

If we substitute the corresponding Laurent series for $g_1(z)$ and $f_1(z)$, we obtain a general representation of the form

$$w = D_n\left[\sum_{-\infty}^{\infty} a_\lambda(z-z_o)^\lambda\right] + D_m^*\overline{\left[\sum_{-\infty}^{\infty} b_\lambda(z-z_o)^\lambda\right]} +$$

$$+ \sum_{k=1}^{n} \frac{(-1)^{n-k}(n+m-k)!}{k!(n-k)!\omega^{n-k}} \sum_{s=0}^{k-1} \binom{k}{s} R^s q(\varphi) R^{k-1-s}\left[\frac{1}{(z-z_o)\varphi'}\right] +$$

$$+ \sum_{k=1}^{m} \frac{(-1)^{m-k}(n+m-k)!}{k!(m-k)!\omega^{n-k}} \sum_{s=0}^{k-1} \binom{k}{s} \frac{m!}{n!}(-1)^{n+m} S^s q(-\bar{\psi}) S^{k-1-s}\left[\frac{1}{(z-z_o)\psi'}\right] +$$

$$+ 2D_n q \log|z-z_o|.$$

In particular we point to a special case. By $S^{n+m+1}\bar{f} \equiv 0$ we get $p_2(\psi) \equiv 0$ and therefore $p_1(\varphi) \equiv 0$. By this we obtain the following theorem for the solutions of (56) which can be represented by only one generator.

Theorem 8

Let w be a solution of (56) in $\dot{U}(z_o)$ which has an isolated singularity at z_o and can be represented by only one generator $g(z)$.
Then, w may be represented in $\dot{U}(z_o)$ by

$$w = D_n g ,$$

where

$$g(z) = \sum_{-\infty}^{\infty} a_\lambda(z-z_o)^\lambda$$

is a holomorphic and unique function in $U(z_o)$.

c) The special cases $w_{z\bar{z}} - n(n+1)Gw = 0$ and $(1+\varepsilon z\bar{z})^2 w_{z\bar{z}} + \varepsilon n(n+1)w = 0$

Setting m = n in (56) the summand with the first derivative vanishes and we get the differential equation

$$w_{z\bar{z}} - n(n+1)Gw = 0, \quad n \in \mathbb{N}, \tag{67}$$

with

$$G = \frac{\varphi'\overline{\psi'}}{(\varphi+\bar{\psi})^2},$$

where G is the solution of (17)

$$(\log G)_{z\bar{z}} + G = 0.$$

Here, the functions $\varphi(z)$ and $\psi(z)$ may be holomorphic or meromorphic, providing that they satisfy certain conditions with respect to their poles (cf. Theorem I, (20)). The corresponding representation theorems for the solutions of (67) in simply connected domains and in the neighbourhood of isolated singularities were proved in [10], [11], and [25]. If G denotes a real-valued solution of (17) (cf. Theorem 1,c and d), also the differential equation (67) has real-valued solutions.

Since the solutions of (67) are of special interest in view of their function theoretic properties and their importance in connection with applications [4)], in the following we formulate the corresponding representation theorems; for the rest we refer the reader to [10], [11], and [25]).

We denote by $M_{2n}(\varphi,D)$ respectively $M_{2n}(\psi,D)$ the set of functions which are holomorphic or meromorphic in D and have only a finite number of

[4)] Considering circulation-free subsonic gas flow, one is led, for example, to differential equations of type (67) (cf. e.g.[85]).

poles which appear at most in such points in which the functions $\varphi(z)$ respectively $\psi(z)$ have poles too. By H_n respectively H_n^* we denote the differential operators

$$(68) \qquad H_n = \sum_{k=0}^{n} \frac{A_k^n}{\omega^{n-k}} R^k,$$

$$(69) \qquad H_n^* = \sum_{k=0}^{n} \frac{A_k^n}{\omega^{n-k}} S^k$$

with

$$\omega = \varphi + \overline{\psi}, \qquad A_k^n = \frac{(-1)^{n-k}(2n-k)!}{k!(n-k)!},$$

$$R = \frac{1}{\varphi'} \frac{\partial}{\partial z}, \qquad S = \frac{1}{\overline{\psi'}} \frac{\partial}{\partial \bar{z}}.$$

Moreover, we use the operators

$$P = \omega^2 R \quad \text{and} \quad Q = \omega^2 S .$$

Theorem 9

Let $\varphi(z)$ and $\psi(z)$ be holomorphic or meromorphic functions in D which satisfy the conditions (20).

a) For every solution w of (67)

$$\frac{(\varphi+\overline{\psi})^2}{\varphi'\overline{\psi'}} w_{z\bar{z}} - n(n+1)w = 0, \qquad n \in \mathbb{N},$$

defined in D, there exist two functions

$$g(z) \in M_{2n}(\varphi,D) \quad \text{and} \quad h(z) \in M_{2n}(\psi,D),$$

such that

$$(70) \qquad w = H_n g + H_n^* \bar{h}$$

$$= \omega^{n+1}\left[R^n\left[\frac{g(z)}{\omega^{n+1}}\right] + S^n \left[\overline{\frac{h(z)}{\omega^{n+1}}}\right]\right]$$

$$= \omega^{n+1} R^n S^n \left[\frac{u}{\omega}\right]$$

with

$$u(z,\bar{z}) = \frac{(-1)^n}{n!}[g(z) + \overline{h(z)}] .$$

b) Conversely, for each pair of functions

$$g(z) \in M_{2n}(\varphi,D) \quad \text{and} \quad h(z) \in M_{2n}(\psi,D)$$

(70) represents a solution of (67) in D.

c) For every given solution w the functions $R^{2n+1}g$ and $S^{2n+1}\bar{h}$ are uniquely determined by

$$(71) \qquad R^{2n+1}g = \frac{P^{n+1}w}{\omega^{2n+2}} , \qquad S^{2n+1}\bar{h} = \frac{Q^{n+1}w}{\omega^{2n+2}} .$$

In this case the generators g(z) and h(z) are not uniquely determined. We get the most general pair of generators $\tilde{g}(z)$ and $\tilde{h}(z)$ by

$$(72) \qquad \tilde{g}(z) = g(z) + \sum_{\mu=0}^{2n} a_\mu \varphi^\mu ,$$

$$(73) \qquad \tilde{h}(z) = h(z) - \sum_{\mu=0}^{2n} (-1)^\mu \, \overline{a_\mu} \, \psi^\mu , \qquad a_\mu \in \mathbb{C}.$$

d) For every solution w of (67) in D which can be represented by only one generator g(z) or h(z) this function is uniquely determined by

$$(74) \qquad g(z) = \frac{Q^n w}{(2n)!} ,$$

(75) $$h(z) = \frac{\overline{P^n w}}{(2n)!} .$$

If the coefficient G in (67) is a real-valued solution of the differential equation (17), we have the following result.

Theorem 10

Let D be a simply connected domain of the complex plane. Let f(z) be a holomorphic or meromorphic function in D which satisfies the conditions (22).

a) For every solution w of the differential equation

(76) $$(1+\varepsilon f\bar{f})^2 w_{z\bar{z}} + \varepsilon n(n+1) f' \overline{f'} w = 0, \qquad n \in \mathbb{N}, \qquad \varepsilon = \pm 1,$$

defined in D, there exist two functions

$$g(z),\ h(z) \in M_{2n}(f,D),$$

such that

(77) $$w = H_n g + \overline{H_n h}$$

with

$$H_n = \sum_{k=0}^{n} \frac{A_k^n}{\omega^{n-k}} R^k, \qquad \omega = \frac{1+\varepsilon f\bar{f}}{\varepsilon \bar{f}}, \qquad R = \frac{1}{f'} \frac{\partial}{\partial z} .$$

b) Conversely, for each pair of functions

$$g(z),\ h(z) \in M_{2n}(f,D)$$

(77) represents a solution of (76) in D.

c) For every given solution w of (76) the functions

(78) $$R^{2n+1} g = \frac{P^{n+1} w}{\omega^{2n+2}}, \qquad R^{2n+1} h = \frac{P^{n+1} \bar{w}}{\omega^{2n+2}}$$

are uniquely determined. In this case the generators $g(z)$ and $h(z)$ are only determined up to a polynomial $p(f)$ of degree 2n. We obtain the most general pair of generators $\tilde{g}(z)$ and $\tilde{h}(z)$ by

$$\tilde{g}(z) = g(z) + p(f), \quad \tilde{h}(z) = h(z) - (-\varepsilon)^n f^{2n} \overline{p\left(\frac{-\varepsilon}{\bar{f}}\right)} \quad . \tag{79}$$

d) For every solution of (76) which can be represented in the form

$$w = H_n g \quad \text{or} \quad w = \overline{H_n h}$$

the generator is uniquely determined by

$$g(z) = \frac{\overline{P^n \bar{w}}}{(2n)!}, \quad h(z) = \frac{\overline{P^n w}}{(2n)!} \ . \tag{80}$$

e) For every real-valued solution of (76) in D there exists a function $g(z) \in M_{2n}(f,D)$, such that

$$w = H_n g + \overline{H_n g} \ . \tag{81}$$

f) Conversely, for each function $g(z) \in M_{2n}(f,D)$ (81) represents a real-valued solution of (76) in D.

g) For every given real-valued solution w of (76) the function

$$R^{2n+1} g = \frac{P^{n+1} w}{\omega^{2n+2}}$$

is uniquely determined. In this case the generator $g(z)$ is only determined up to a polinomial $p(f)$ of degree 2n. We get the most general generator $\tilde{g}(z)$ by

$$\tilde{g}(z) = g(z) + p(f) \tag{82}$$

with

$$p(f) + (-\varepsilon)^n f^{2n} \overline{p\left(\frac{-\varepsilon}{\bar{f}}\right)} = 0 \ . \tag{83}$$

It is also possible to get corresponding representations for the solutions of inhomogeneous differential equations

$$(84)\qquad \frac{(\varphi+\overline{\psi})^2}{\varphi'\overline{\psi'}}\, w_{z\overline{z}} - n(n+1)w = \Phi(z,\overline{z})$$

if the function $\Phi(z,\overline{z})$ satisfies certain conditions.
First, we suppose that $\Phi = \Phi_k(z,\overline{z})$ denotes a solution of the homogeneous differential equation

$$(85)\qquad \frac{(\varphi+\overline{\psi})^2}{\varphi'\overline{\psi'}}\, \Phi_{k,z\overline{z}} - k(k+1)\Phi_k = 0, \quad k \in \mathbb{N}, \quad k \neq n,$$

defined in D. For a particular solution of (84) we set

$$w = \lambda\Phi_k(z,\overline{z}), \qquad \lambda \in \mathbb{R}.$$

If we substitute into (84), it follows by (85)

$$\Phi_k \left\{ \lambda[k(k+1) - n(n+1)] - 1\right\} = 0.$$

Therefore, we get a particular solution with

$$\lambda = \frac{1}{k(k+1) - n(n+1)} .$$

Similary, we obtain a particular solution if the term $\Phi(z,\overline{z})$ is a sum of solutions of homogeneous differential equations of the type (85). The method fails only for $k = n$, i.e. in case of resonance.

Theorem 11

Let $\Phi_k(z,\overline{z})$, $k = 0,1, \ldots, m$, $k \neq n$, be solutions of the homogeneous differential equations (85)

$$\frac{(\varphi+\overline{\psi})^2}{\varphi'\overline{\psi'}}\, \Phi_{k,z\overline{z}} - k(k+1)\Phi_k = 0$$

defined in D. Then

(86) $$w = \sum_{\substack{k=0 \\ k \neq n}}^{m} \frac{1}{k(k+1)-n(n+1)} \Phi_k(z,\bar{z})$$

represents a particular solution of the inhomogeneous differential equation

(87) $$\frac{(\varphi+\bar{\psi})^2}{\varphi'\overline{\psi'}} w_{z\bar{z}} - n(n+1)w = \sum_{\substack{k=0 \\ k \neq n}}^{m} \Phi_k(z,\bar{z})$$

in D.

If a differential equation of the form (84) is given, in view of the applications of these results the question arises whether it is possible to represent the term $\Phi(z,\bar{z})$ as a sum of functions $\Phi_k(z,\bar{z})$. Here we are led to the differential equation

(88) $$Q^{m+1}\left[\frac{P^{m+1}\Phi}{\omega^{2m+2}}\right] = 0$$

which was treated for the case

$$P = \omega^2 r, \quad Q = \omega^2 s, \quad \omega = 1 + \varepsilon z\bar{z}, \quad \varepsilon = \pm 1,$$

in [7].

If the inhomogeneous term Φ in (84) has the form

(89) $$\Phi = \omega^{2+n-k}\Phi_{1,k}, \quad k \in \mathbb{N}, \quad k \neq n+1,$$

with

$$\Phi_{1,k} = H_k g_{1,k} + H_k^* \overline{h_{1,k}}, \quad g_{1,k}(z), \quad h_{1,k}(z) \in H(D),$$

we get a particular solution of (84) by

$$w = \frac{\omega^{1+n-k}}{1+n-k} [H_{k-1} g_{1,k} + H_{k-1}^* \overline{h_{1,k}}] .$$

If the term Φ in (84) has the form

$$\Phi = \omega^{1-n-k}\Phi_{2,k}\,, \quad k \in \mathbb{N}, \tag{90}$$

with

$$\Phi_{2,k} = H_k g_{2,k} + H_k^* \overline{h_{2,k}}\,, \quad g_{2,k}(z), \quad h_{2,k}(z) \in H(D),$$

we obtain a particular solution by

$$w = -\frac{\omega^{-n-k}}{n+k}\left[H_{k-1}g_{2,k} + H_{k-1}^*\overline{h_{2,k}}\right].$$

Similarly we get particular solutions if the inhomogeneous term represents a sum of terms of the form (89) and (90), where the functions $\Phi_{j,k}$, $j = 1,2$, denotes solutions of the homogeneous differential equation

$$\frac{(\varphi+\overline{\varphi})^2}{\varphi'\overline{\varphi'}}\frac{\partial^2}{\partial z\partial\overline{z}}\Phi_{j,k} - k(k+1)\Phi_{j,k} = 0 \tag{91}$$

in D.

Theorem 12

Let $\Phi_{j,k}$, $j = 1,2$, $k \in \mathbb{N}$, be solutions of (91) defined in D with the representation

$$\Phi_{j,k} = H_k g_{j,k} + H_k^*\overline{h_{j,k}}\,,$$

$$g_{j,k}(z),\ h_{j,k}(z) \in H(D).$$

Then,

$$w = \sum_{\substack{k=1\\ k\neq n+1}}^{m_1} \frac{\omega^{1+n-k}}{1+n-k}\left[H_{k-1}g_{1,k} + H_{k-1}^*\overline{h_{1,k}}\right]$$

$$- \sum_{k=1}^{m_2} \frac{\omega^{-n-k}}{n+k}\left[H_{k-1}g_{2,k} + H_{k-1}^*\overline{h_{2,k}}\right]$$

is a particular solution of the inhomogeneous differential equation

$$\frac{(\varphi+\bar{\psi})^2}{\varphi'\overline{\psi'}} w_{z\bar{z}} - n(n+1)w = \sum_{\substack{k=1 \\ k\neq n+1}}^{m_1} \omega^{2+n-k}\Phi_{1,k} + \sum_{k=1}^{m_2} \omega^{1-n-k}\Phi_{2,k}$$

in D.

Setting $f(z) = z$ in (76), we get the differential equation

$$(92) \qquad (1+\varepsilon z\bar{z})^2 w_{z\bar{z}} + \varepsilon n(n+1)w = 0, \quad n \in \mathbb{N}, \quad \varepsilon = \pm 1,$$

which is closely related to the Laplace and wave equations and results from certain transformations and following separation. Equation (92) has been investigated by many mathematicians. In (92), apart from w, the second Beltrami operator appears; therefore, this differential equation is invariant under all rotations of the sphere in the case $\varepsilon = 1$ and under all automorphisms of the unit disk in the case $\varepsilon = -1$. For this reason, in particular in the case $\varepsilon = -1$, the construction of automorphic solutions is of special interest. Here, for example, E. Peschl succeeded in construction such solutions by means of absolut differential invariants.
Proceeding from an arbitrary meromorphic function $f(z)$, we consider the following terms for $z\bar{z} < 1$:

$$(93) \qquad \alpha_2 = \frac{f''}{2f'^2} - \frac{\bar{z}}{f'(1-z\bar{z})},$$

$$(94) \qquad \beta_3 = \frac{1}{6f'^2}[f]_z, \text{ [5)]}$$

$$(95) \qquad \beta_m = \frac{1}{mf'}\frac{\partial}{\partial z}\beta_{m-1}, \quad m \geq 4.$$

5) $[f]_z$ denotes the Schwarzian derivative (Schwarzian differential invariant)

$$[f]_z = \left(\frac{f''}{f'}\right)' - \frac{1}{2}\left(\frac{f''}{f'}\right)^2.$$

The lower index of these quantities gives the order of the highest derivative of f(z) which arises. α_2 is a function of z and $\bar{z}$, whereas the quantities β_k only depends on z. If we denote the group of automorphisms of the unit disk by

$$L_{-1}: z = \rho(\zeta) = \eta \frac{\zeta + a}{1 + \bar{a}\zeta}, \qquad |\eta| = 1, \qquad |a| < 1,$$

and use $I\{f(z)\}$ for a differential term of the form (93) - (95), it follows

$$\underset{\zeta}{I}\{[f(z)]_{z=\rho(\zeta)}\} = \left[\underset{z}{I}\{f(z)\}\right]_{z=\rho(\zeta)} .$$

To say, the quantities (93) - (95) represent absolut differential invariants under all transformations of the group L_{-1}. Then, by

$$(96) \qquad w = \sum_{\nu=0}^{n} B_\nu(\beta_3, \ldots, \beta_{k+n-\nu})\alpha_2^\nu , \qquad k \geq 3,$$

we obtain a solution of the differential equation

$$(97) \qquad (1-z\bar{z})^2 w_{z\bar{z}} - n(n+1)w = 0, \qquad n \in \mathbb{N},$$

if B_n is an arbitrary function of the invariants $\beta_3, \ldots, \beta_k$ and

$$B_\nu = \frac{\nu+1}{(\nu+1+n)(\nu-n)} \left[\sum_{\lambda \geq 3} (\lambda+1)\beta_{\lambda+1} B_{\nu+1,\beta_\lambda} + 3(\nu+2)\beta_3 B_{\nu+2} \right]$$

$$\text{for } 0 \leq \nu \leq n-1 \quad \text{with} \quad B_{n+1} \equiv 0.$$

Therefore, also w is a differential operator acting on f(z). By

$$w = I\{f(z)\}$$

also here we have

$$\underset{\zeta}{I}\{[f(z)]_{z=\rho(\zeta)}\} = \left[\underset{z}{I}\{f(z)\}\right]_{z=\rho(\zeta)}$$

for $z = \rho(\zeta) \in L_{-1}$.

Let L_{-1}^{*} be a properly discontinuous subgroup of L_{-1}. If now f(z) is an automorphic function with respect to L_{-1}^{*}, first, it follows

$$[f(z)]_{z=\rho^{*}(\zeta)} = f(\zeta)$$

and

$$\underset{\zeta}{I}\left\{f(\zeta)\right\} = \left[\underset{z}{I}\left\{f(z)\right\}\right]_{z=\rho^{*}(\zeta)}$$

for each transformation $\rho^{*}(\zeta) \in L_{-1}^{*}$, and therefore

$$[w(z,\bar{z})]_{z=\rho^{*}(\zeta)} = w(\zeta,\bar{\zeta})$$

for a solution (96). That is, $w(z,\bar{z})$ represents a solution of (97) which is automorphic with respect to L_{-1}^{*}.

Among other things this result of E.Peschl gave rise to deduce a representation by differential operators for all solutions of (92) which are defined in simply connected domains of the unit disk ($\varepsilon = -1$) and in the complex plane ($\varepsilon = +1$) respectively. Moreover, the function theoretic properties of the solutions of (92) were investigated by several mathematicians. In this way a function theory was developed associated to (92). These problems are treated in the subsequent contribution of St.Ruscheweyh. Therefore, we summarize here some important results concerning the representation of the solutions of (92) and use the following notations:

$$E_n = \sum_{\nu=0}^{n} A_\nu^n \bar{\tau}^{n-\nu} \frac{d^\nu}{dz^\nu}, \tag{98}$$

$$A_\nu^n = \frac{(-\varepsilon)^{n-\nu}(2n-\nu)!}{\nu!(n-\nu)!}, \qquad \tau = \frac{z}{1+\varepsilon z\bar{z}}, \tag{99}$$

$$D_\varepsilon = (1+\varepsilon z\bar{z})^2 \frac{\partial}{\partial z}. \tag{100}$$

We denote by D a simply connected domain of the unit disk ($\varepsilon = -1$) respectively the complex plane ($\varepsilon = +1$).

Theorem 13

a) For every solution w of the differential equation (92)

$$(1+\varepsilon z\bar{z})^2 w_{z\bar{z}} + \varepsilon n(n+1)w = 0, \qquad \varepsilon = \pm 1, \qquad n \in \mathbb{N},$$

defined in D, there exist two functions $g(z), f(z) \in H(D)$, such that

$$(101) \qquad w = E_n g + \overline{E_n f}$$

$$= (1+\varepsilon z\bar{z})^{n+1}\left[\frac{\partial^n}{\partial z^n}\left[\frac{g(z)}{(1+\varepsilon z\bar{z})^{n+1}}\right] + \frac{\partial^n}{\partial \bar{z}^n}\left[\frac{\overline{f(z)}}{(1+\varepsilon z\bar{z})^{n+1}}\right]\right].$$

b) Conversely, for arbitrary functions $g(z), f(z) \in H(D)$ (101) represents a solution of (92) in D.

c) For every given solution w of (92) the functions $g^{(2n+1)}(z)$ and $f^{(2n+1)}(z)$ are uniquely determined by

$$(102) \qquad g^{(2n+1)}(z) = \frac{D_\varepsilon^{n+1} w}{(1+\varepsilon z\bar{z})^{2n+2}},$$

$$(103) \qquad f^{(2n+1)}(z) = \frac{D_\varepsilon^{n+1} \bar{w}}{(1+\varepsilon z\bar{z})^{2n+2}}.$$

In this case the generators g(z) and f(z) are only determined up to a polynomial p(z) of degree 2n. We obtain the most general pair of generators $\tilde{g}(z)$ and $\tilde{f}(z)$ by

$$(104) \qquad \tilde{g}(z) = g(z) + p(z), \qquad \tilde{f}(z) = f(z) - (-\varepsilon)^n z^{2n}\overline{p\left(\frac{-\varepsilon}{\bar{z}}\right)}.$$

d) For every solution w which can be represented by only one generator g(z) or f(z) this function is uniquely determined by

$$(105) \qquad g(z) = \frac{(-\varepsilon)^n}{(2n)!}\overline{D_\varepsilon^n \bar{w}},$$

(106) $$f(z) = \frac{(-\varepsilon)^n}{(2n)!} \overline{D_\varepsilon^n w} .$$

e) For every real-valued solution w of (92) in D there exists a function $g(z) \in H(D)$, such that

(107) $$w = E_n g + \overline{E_n g} .$$

f) Conversely, for each function $g(z) \in H(D)$ (107) represents a real-valued solution of (92) in D.

Theorem 14

a) For every solution of the differential equation

(108) $$(1+z\bar{z})^2 w_{z\bar{z}} + n(n+1)w = 0, \qquad n \in \mathbb{N},$$

which is defined on the whole Riemann number sphere there exist 2n+1 constants $c_\mu \in \mathbb{C}$, $\mu = 0,1, \ldots, 2n$, such that

(109) $$w = E_n g$$

with

(110) $$g(z) = \sum_{\mu=0}^{2n} c_\mu z^\mu .$$

b) Conversely, for arbitrary constants $c_\mu \in \mathbb{C}$ (109) represents a solution of (108) defined on the whole Riemann number sphere.

c) For every given solution of (108) defined on the whole Riemann number sphere the function g(z) is uniquely determined by

(111) $$g(z) = \frac{(-1)^n}{(2n)!} \overline{D_{+1}^n \bar{w}} .$$

The behavior of the solutions of (92) in the neighbourhood of isolated singularities was investigated in [26]. Some of these results can be obtained by specializing the functions $\varphi(z)$ and $\psi(z)$ in Theorem 7 and

Theorem 8. Moreover, in [26] we find an investigation of the isolated singularities with a logarithmic principal term of the asymptotic expansion. In this context we refer the reader in particular to the Theorems 5, 8, 10, and 11 in [26]. The branched solutions of (92) were investigated by L.Reich in [94].
In [8] one can find a representation of the solutions of the inhomogeneous differential equation

$$(112) \qquad (1+\varepsilon z\bar{z})^2 w_{z\bar{z}} + \varepsilon n(n+1)w = \Phi(z,\bar{z}), \qquad \varepsilon = \pm 1, \qquad n \in \mathbb{N},$$

with

$$\Phi(z,\bar{z}) = \sum_{j=1}^{n} \left[\left[\frac{\bar{z}}{1+\varepsilon z\bar{z}} \right]^{n-j} \varphi_j(z) + \left[\frac{z}{1+\varepsilon z\bar{z}} \right]^{n-j} \overline{\psi_j(z)} \right],$$

$$\varphi_j(z),\ \psi_j(z) \in H(D),$$

in simply connected domains and in the neighbourhood of isolated singularities.

3) Differential operators on solutions of differential equations of the form $w_{z\bar{z}} + Aw_{\bar{z}} + Bw = 0$

In Theorem 6 and Theorem 9 representations of solutions are given in which an arbitrary solution of the differential equation

$$h_{z\bar{z}} = 0$$

appears as generator. This result suggested the idea to use solutions of the differential equation

$$h_{z\bar{z}} + B(z,\bar{z})h = 0 \tag{113}$$

as generators. [6)]

By D we denote again a simply connected domain of the complex plane. Let $\gamma(z)$ and $\delta(z)$ be holomorphic or meromorphic functions in D which satisfy the following conditions ($\eta = \gamma + \bar{\delta}$):

(114)
- (i) $\gamma(z)$ and $\delta(z)$ have only a finite number of poles in D.
- (ii) $\gamma(z)$ and $\delta(z)$ have no common poles in D.
- (iii) $\eta\gamma'\delta' \neq 0$ in D.

Again we use the operators

$$r = \frac{\partial}{\partial z}, \quad s = \frac{\partial}{\partial \bar{z}}$$

and set

$$K = \frac{1}{\gamma'} r + \frac{1}{\overline{\delta'}} s, \quad w_o = \frac{h}{\eta}. \tag{115}$$

Then, w_o satisfies the differential equation

---

6) Similar investigations can be found in [14], where solutions of partial differential equations arise as generators which are defined in polydomains of the space $\mathbb{C}^m$.

$$\frac{\eta}{\gamma'\delta'} rsw_o + Kw_o + \frac{B}{\gamma'\delta'} w_o = 0. \tag{116}$$

If we apply the operator

$$K - \frac{2}{\eta} \tag{117}$$

to (116), by

$$K(rs) = (rs)K - \left[\left(\frac{1}{\gamma'}\right)' + \overline{\left(\frac{1}{\delta'}\right)} \right] rs \tag{118}$$

with $Kw_o = w_1$ it follows the differential equation

$$\frac{\eta}{\gamma'\delta'} rsw_1 + Kw_1 + \frac{B\eta}{\gamma'\delta'} w_1 = \frac{2}{\eta} w_1 ,$$

if B satisfies the differential equation

$$KB + \left[\left(\frac{1}{\gamma'}\right)' + \overline{\left(\frac{1}{\delta'}\right)'} \right] B = 0. \tag{119}$$

Consequently,

$$B = \gamma'\overline{\delta'}\,\Psi(\vartheta), \qquad \vartheta = \gamma - \overline{\delta} , \tag{120}$$

where $\Psi(\vartheta)$ denotes an arbitrary continuously differentiable function of ϑ . Setting

$$w_k = \left[K - \frac{2(k-1)}{\eta} \right] w_{k-1} , \qquad k = 1,2, \ldots, n,$$

by induction it follows that w_n is a solution of the differential equation

$$\frac{\eta}{\gamma'\overline{\delta'}} rsw_n + Kw_n + \frac{B}{\gamma'\overline{\delta'}} w_n = \frac{n(n+1)}{\eta} w_n . \tag{121}$$

By the transformation

(122) $$w_n = \frac{w}{\eta}$$

we get

(123) $$w_{z\bar{z}} + \left[B - \frac{n(n+1)\gamma'\overline{\delta'}}{\eta^2}\right] w = 0, \qquad n \in \mathbb{N}.$$

Since the coefficient B is a solution of (119), it follows that Kh is a solution of (113) if h is a solution of this differential equation. Therefore, we may expect that the above-derived solution

(124) $$w = \eta\left[K - \frac{2(n-1)}{\eta}\right] \dots \left[K - \frac{2}{\eta}\right] K \frac{h}{\eta}$$

of (123) can be represented in a simpler way. In fact, this is possible. We obtain

(125) $$w = \sum_{k=0}^{n} \frac{(-1)^{n-k}(2n-k)!}{k!(n-k)!} \frac{K^k h}{\eta^{n-k}},$$

as can be proved by induction on n.

Theorem 15

Let $\gamma(z)$ and $\delta(z)$ be holomorphic or meromorphic functions in D which satisfy the conditions (114). Let $h(z,\bar{z})$ be a solution of the differential equation

$$h_{z\bar{z}} + Bh = 0$$

defined in D, where

$$B(z,\bar{z}) = \gamma'\overline{\delta'}\,\Psi(\vartheta), \qquad \vartheta = \gamma - \bar{\delta},$$

denotes a continuously differentiable function in D.
Then,

(126) $$w = \eta\left[K - \frac{2(n-1)}{\eta}\right] \dots \left[K - \frac{2}{\eta}\right] K \frac{h}{\eta} = \sum_{k=0}^{n} \frac{(-1)^{n-k}(2n-k)!}{k!(n-k)!} \frac{K^k h}{\eta^{n-k}}$$

with

$$K = \frac{1}{\gamma'}\frac{\partial}{\partial z} + \frac{1}{\delta'}\frac{\partial}{\partial \bar{z}} \quad \text{and} \quad \eta = \gamma + \bar{\delta}$$

represents a solution of the differential equation

$$(127) \qquad w_{z\bar{z}} + \gamma'\overline{\delta'}\left[\psi(\vartheta) - \frac{n(n+1)}{\eta^2}\right]w = 0$$

in D.

As the case may be, it depends on the form of the function B in (113) that we have to use integral operators for the representation of the generator $h(z,\bar{z})$. In this context the question is of interest whether there are differential equations of the type (123) for which the above-derived solutions can be represented only by differential operators.

Considering Theorem 9, we obtain such a differential equation, for example, if we set

$$B(z,\bar{z}) = \frac{-m(m+1)\varphi'\overline{\varphi'}}{(\varphi+\bar{\varphi})^2}, \quad m \in \mathbb{N}_0 .$$

In this case equation (119) is satisfied, for instance, by

$$(128) \qquad \gamma = C_1 - \frac{1}{\varphi+C_2}, \quad \bar{\delta} = \frac{1}{\bar{\psi} - C_2}, \quad C_1, C_2 \in \mathbb{C},$$

and by Theorem 15 we get representations of solutions of the differential equation

$$(129) \qquad w_{z\bar{z}} - \varphi'\overline{\varphi'}\left[\frac{m(m+1)}{(\varphi+\bar{\varphi})^2} - \frac{n(n+1)}{[C_1(\varphi+C_2)(\bar{\psi}-C_2)+\varphi-\bar{\psi}+2C_2]^2}\right]w = 0 . \quad {}^{7)}$$

---

7) If we formally substitute z_1 for z and z_2 for $\bar{z}$ into (129), with $C_1 = C_2 = 0$, $\varphi = z_1$, $\bar{\psi} = z_2$ we get

$$(130) \qquad w_{z_1 z_2} - \left[\frac{m(m+1)}{(z_1+z_2)^2} - \frac{n(n+1)}{(z_1-z_2)^2}\right]w = 0.$$

For this equation G. Jank [69] derived representations of solutions by using an other method.

In view of a generalized Darboux equation which is considered in Chapter I,5 we will point to another possibility of the representation of solutions. To the differential equation (116) we apply the operator

$$K^* + \frac{\gamma - \overline{\delta}}{\eta}$$

with

$$K^* = -\frac{\gamma}{\gamma'}\, r + \frac{\overline{\delta}}{\overline{\delta'}}\, s \, .$$

Considering

$$K^*(rs) = (rs)K^* - \left[\left(-\frac{\gamma}{\gamma'}\right)' + \overline{\left(\frac{\delta}{\delta'}\right)'} \right] rs \tag{131}$$

and

$$K^*K = KK^* - \left[-\frac{1}{\gamma'}\, r + \frac{1}{\overline{\delta'}}\, s \right] \tag{132}$$

we get

$$\frac{\eta}{\gamma'\overline{\delta'}}\, rsw_1 + Kw_1 + \frac{B\eta}{\gamma'\overline{\delta'}}\, w_1 = \frac{2}{\eta}\, w_1 \, ,$$

with $K^*w_o = w_1$ if $B(z,\overline{z})$ is a solution of

$$K^*B + \left[\left(-\frac{\gamma}{\gamma'}\right)' + \overline{\left(\frac{\delta}{\delta'}\right)'} \right] B = 0. \tag{133}$$

Setting

$$w_k = \left[K^* + (k-1)\frac{\gamma-\overline{\delta}}{\eta} \right] w_{k-1} \, , \quad k = 1, \ldots, n,$$

it follows by induction that w_n is a solution of the differential equation (121), where B now denotes a solution of (133). In this case we see by (131) that K^*h is a solution of (113) if h is a solution of this differential equation. By induction it follows that $w = \eta\, w_n$ can be represented in the form

$$w = \sum_{k=0}^{n} \frac{(2n-k)!}{k!(n-k)!} \left(\frac{\gamma}{\eta}\right)^{n-k} (K^*-n) \ldots (K^*-n+k-1)h.$$

Theorem 16

Let $\gamma(z)$ and $\delta(z)$ be holomorphic or meromorphic functions in D which satisfy the conditions (114). Let $h(z,\bar{z})$ be a solution of the differential equation

$$h_{z\bar{z}} + B(z,\bar{z})h = 0,$$

defined in D, where B is a solution of

$$K^*B + \left[\left(-\frac{\gamma}{\gamma'}\right)' + \overline{\left(\frac{\delta}{\delta'}\right)'}\right] B = 0$$

in D with

$$K^* = -\frac{\gamma}{\gamma'}\frac{\partial}{\partial z} + \frac{\bar{\delta}}{\bar{\delta}'}\frac{\partial}{\partial \bar{z}} .$$

Then,

$$w = \eta\left[K^* + (n-1)\frac{\gamma-\bar{\delta}}{\eta}\right] \ldots \left[K^* + \frac{\gamma-\bar{\delta}}{\eta}\right] K^* \frac{h}{\eta}$$

$$= \sum_{k=0}^{n} \frac{(2n-k)!}{k!(n-k)!} \left(\frac{\gamma}{\eta}\right)^{n-k} (K^*-n)_k h$$

with

$$\eta = \gamma(z) + \overline{\delta(z)} ,$$

$$(K^*-n)_k = (K^*-n)(K^*-n+1) \ldots (K^*-n+k-1) \quad \text{for} \quad k \in \mathbb{N} ,$$

$$(K^*-n)_0 h = h,$$

represents a solution of the differential equation

$$w_{z\bar{z}} + \left[B - \frac{n(n+1)\gamma'\bar{\delta}'}{\eta^2}\right] w = 0$$

in D.

The above results suggest to ask for differential operators of the form

$$L = a_1 r + a_2 s + a_3 \tag{134}$$

which map a solution of the differential equation

$$w_{z\bar{z}} + Bw = 0 \tag{135}$$

onto a solution

$$v = Lw \tag{136}$$

of the differential equation

$$v_{z\bar{z}} + B^* v = 0. \tag{137}$$

If we substitute (136) into (137), it follows by (135)

$$\left\{\begin{array}{l} a_1 = a_1(z), \quad a_2 = a_2(\bar{z}) \\ a_1 B^* = a_1 B - s a_3, \\ a_2 B^* = a_2 B - r a_3, \\ r(a_1 B) + s(a_2 B) + a_3(B - B^*) = r s a_3. \end{array}\right. \tag{138}$$

Setting

$$a_1(z) = \frac{1}{\gamma'(z)}, \quad a_2(\bar{z}) = \frac{1}{\overline{\delta'(z)}},$$

we obtain

$$\overline{\delta'} r a_3 = \gamma' s a_3,$$

$$a_3(z,\bar{z}) = \Phi(\eta) \quad \text{with} \quad \eta = \gamma(z) + \overline{\delta(z)}$$

and

$$B^* = B - \gamma' \overline{\delta'} \, \Phi' . \tag{139}$$

Then, by (138) $B(z,\bar{z})$ satisfies the differential equation

$$r \frac{B}{\gamma'} + s \frac{B}{\overline{\delta'}} = \gamma'\overline{\delta'}(\Phi''-\Phi\Phi') .$$

Setting here

$$\Phi'' - \Phi\Phi' = 2\,\Psi_1'(\eta) ,$$

it follows

$$B = \gamma'\overline{\delta'}[\Psi_1(\eta) + \Psi_2(\vartheta)], \tag{140}$$

where $\Psi_2(\vartheta)$ is an arbitrary function of $\vartheta = \gamma - \bar{\delta}$.

Theorem 17

Let w be a solution of the differential equation

$$w_{z\bar{z}} + \gamma'\overline{\delta'}[\Psi_1(\eta) + \Psi_2(\vartheta)]w = 0, \quad \eta = \gamma + \bar{\delta}, \quad \vartheta = \gamma - \bar{\delta}.$$

Then, by $v = Lw$ with

$$L = \frac{1}{\gamma'}\frac{\partial}{\partial z} + \frac{1}{\overline{\delta'}}\frac{\partial}{\partial \bar{z}} + \Phi \tag{141}$$

we get a solution of the differential equation

$$v_{z\bar{z}} + \gamma'\overline{\delta'}[\Psi_1(\eta) + \Psi_2(\vartheta) - \Phi'(\eta)]v = 0,$$

if $\Phi(\eta)$ is a solution of the Riccati equation

$$2\Phi'(\eta) = \Phi^2(\eta) + 4\Psi_1(\eta) + C_1, \quad C_1 \in \mathbb{C}. \tag{142}$$

This result may be applied in different ways. On the one hand it is possible to get new representations for solutions of differential equations treated above. On the other hand by the solutions of (142) we are led to new differential equations of the form (137) whose solutions can be represented again by differential operators.

For example, if we set

$$C_1 = 0 \text{ and } \psi_1(\eta) = -\frac{n(n+1)}{\eta^2},$$

in (142), it follows

$$2\Phi' = \Phi^2 - \frac{4n(n+1)}{\eta^2}. \tag{143}$$

By means of the particular solution

$$\Phi_1 = \frac{-2(n+1)}{\eta}$$

we obtain the general solution (cf. e.g. [72])

$$\Phi(\eta) = \frac{2C_1 n - 2(n+1)C_2\eta^{2n+1}}{\eta\,[C_1+C_2\eta^{2n+1}]}, \tag{144}$$

with $C_1, C_2 \in \mathbb{C}$ and $(C_1,C_2) \neq (0,0)$. If we set, for instance,

$$\gamma = \varphi, \quad \delta = \psi, \quad \eta = \omega = \varphi + \overline{\psi}, \quad \psi_2(\vartheta) \equiv 0,$$

it follows

$$B = \frac{-n(n+1)}{\omega^2}\varphi'\overline{\psi'}.$$

Thus, it is possible to imploy the results of Theorem 9 concerning the differential equation (67) and we get the following

<u>Theorem 18</u>

Let w be a solution of the differential equation (67)

$$\omega^2 w_{z\overline{z}} - n(n+1)\varphi'\overline{\psi'}\, w = 0.$$

Then, by

$$v = (R+S+\Phi)w \tag{145}$$

we get a solution of the differential equation

$$v_{z\bar{z}} - \left[\frac{n(n+1)}{\omega^2} + \Phi'(\omega)\right]\varphi'\overline{\psi'}v = 0, \tag{146}$$

where $\Phi(\omega)$ is given by (144).

Setting $C_1 = 0$ respectively $C_2 = 0$, we obtain the particular solutions

$$\Phi_1(\omega) = \frac{-2(n+1)}{\omega}, \quad \Phi_2(\omega) = \frac{2n}{\omega}.$$

In these cases (146) becomes a differential equation of the type (67), where the parameter n is to replace by n+1, n-1 respectively. Thus, it is possible to obtain successively the solutions of (67) by means of the differential operators of the type (134). We denote by $F_n(D)$, $n \in \mathbb{N}_o$, the set of the solutions of the differential equation (67)

$$\omega^2 w_{z\bar{z}} - n(n+1)\varphi'\overline{\psi'}w = 0, \quad n \in \mathbb{N}_o,$$

which are defined in D. If we imploy

$$L_m = R + S - \frac{2m}{\omega}, \quad m \in \mathbb{Z}, \tag{147}$$

we get the following

Theorem 19

a) If $w \in F_n(D)$, $n \in \mathbb{N}$, and $u \in F_o(D)$, then

1) $L_{n+1}w \in F_{n+1}(D)$,

2) $L_{-n}w \in F_{n-1}(D)$,

3) $L_nL_{n-1} \dots L_1u \in F_n(D)$.

b) If $w = H_n g + H_n^* \bar{h} \in F_n(D)$, $n \in \mathbb{N}_o$, then

$$v = L_{-(n+1)} L_{n+1} w = L_n L_{-n} w \in F_n(D)$$

with

$$v = H_n(R^2 g) + H_n^*(S^2 \bar{h}).$$

In addition to the operators $L_n L_{-n}$ and $L_{-(n+1)} L_{n+1}$ there also exists a first-order differential operator which maps a solution of the set $F_n(D)$ again onto such a solution. If

$$w = H_n g + H_n^* \bar{h} \in F_n(D),$$

it follows

(148) $$(R-S)w = H_n(Rg) + H_n^*(-S\bar{h}) \in F_n(D).$$

From Theorem 19 it is to be supposed that the solutions of (146) may be built up by solutions of $F_{n+1}(D)$ and $F_{n-1}(D)$. Here, we get the following representation (cf. [28], Theorem 4) as easily can be verified.

Theorem 20

If $w = H_n g + H_n^* \bar{h} \in F_n(D)$, $\omega = \varphi + \bar{\psi}$, and

$$\sigma = \frac{1}{C_2 \omega^{n+1} + C_1 \omega^{-n}}$$

with $(C_1, C_2) \neq (0,0)$, then

(149) $$v = \sigma \left[C_2 \omega^{n+1}(H_{n+1} g + H_{n+1}^* \bar{h}) + C_1 \omega^{-n}(H_{n-1}(R^2 g) + H_{n-1}^*(S^2 \bar{h}))\right]$$

represents a solution of the differential equation (146) in D.

As in case of the differential equation $w_{z\bar{z}} + Bw = 0$ we can deduce similar results in case of the differential equation

$$w_{z\bar{z}} + Aw_{\bar{z}} + Bw = 0$$

by operators of the form (134). In this context here we will point to some properties of the solutions of (56) which correspond to the properties of the solutions of (67) summarized in Theorem 19.

If w is a solution of (56) and if we assume that v is a solution of a differential equation of the same type, we are led to a number of functional-differential-relations by

$$v = Rw + \frac{a}{\omega} w,$$

$$v = \omega Sw + bw,$$

$$v = Rw + cSw + \frac{d}{\omega} w,$$

$$v = \omega^2 Sw$$

which we summarize in the following theorem. We set

$$(150) \qquad L_{1,\nu} = R - \frac{\nu}{\omega}, \quad L_{2,\mu} = \omega S - \mu, \quad \nu,\mu \in \mathbb{Z},$$

and by $F_{m,n}(D)$ we denote the set of solutions of the differential equation (56) in D.

<u>Theorem 21</u>

a) If $w \in F_{m,n}(D)$ with $m,n \in \mathbb{N}$, it follows:

1) $L_{1,m+1} w \in F_{m,n+1}(D)$,

2) $L_{1,-n} w \in F_{m-1,n}(D)$,

3) $L_{2,m+1} w \in F_{m+1,n}(D)$,

4) $L_{2,-n} w \in F_{m,n-1}(D)$,

5) $L_{1,m-n} w \in F_{m-1,n+1}(D)$,

6) $Rw + \frac{m}{n} Sw + \frac{n+m}{\omega} w \in F_{m-1,n-1}(D)$,

7) $Rw + \frac{n+1}{m+1} Sw - \frac{m+n+2}{\omega} w \in F_{m+1,n+1}(D)$,

8) $\omega^2 Sw \in F_{m+1,n-1}(D)$.

b) If $w = D_n g + D_m^* \bar{f} \in F_{m,n}(D)$ with $m,n \in \mathbb{N}_o$, it follows:

1) $(R-S)w = W$,

2) $L_{1,m+1}L_{2,-n}w = nW$,

3) $L_{1,-n}L_{2,m+1}w = -(m+1)W$,

4) $L_{2,m}L_{1,-n}w = -mW$,

5) $L_{2,-(n+1)}L_{1,m+1}w = (n+1)W$,

6) $L_{2,-(n+1)}L_{2,m}L_{1,m-n}w = L_{2,m}L_{2,-(n+1)}L_{1,m-n}w = -m(n+1)W$

with

$$W = D_n(Rg) + D_m^*(-S\bar{f}) \in F_{m,n}(D).$$

4) Linear Bäcklund transformations for differential equations of the type $w_{z\bar{z}} + Bw = 0$

If we know a general representation theorem for the solutions of a differential equation of the form $w_{z\bar{z}} + B(z,\bar{z})w = 0$, it is possible to get representations for the solutions of other differential equations of this type by Bäcklund transformations.[8)]
Let γ be a particular nonvanishing solution of the differential equation

$$w_{z\bar{z}} + B(z,\bar{z})w = 0$$

defined in a simply connected domain D, then

$$w_{z\bar{z}} - \frac{\gamma_{z\bar{z}}}{\gamma} w = 0. \tag{151}$$

By the linear Bäcklund transformation

$$(w-v)_z = \frac{\gamma_z}{\gamma}(w+v), \tag{152a}$$

$$(w+v)_{\bar{z}} = \frac{\gamma_{\bar{z}}}{\gamma}(w-v) \tag{152b}$$

this differential equation is transformed into

$$v_{z\bar{z}} + \left[\frac{\gamma_{z\bar{z}}}{\gamma} - \frac{2\gamma_z\gamma_{\bar{z}}}{\gamma^2}\right] v = 0 \tag{153}$$

as easily can be verified. By integrating (152a) and inserting the result into (152b) we obtain

---

8) With regard to the application of Bäcklund transformations in the theory of ultrashort optical pulses and in the theory of long Josephson junctions we refer the reader to [1,2,82-84], [100] respectively. For an application for hyperbolic differential equations in connection with the infinitesimal deformation of surfaces the reader is directed to [28].

$$(154) \qquad v = -w + \frac{1}{\gamma}\Phi, \quad \Phi_z = 2\gamma w_z, \quad \Phi_{\bar{z}} = 2\gamma_{\bar{z}} w .$$

Proceeding from (152b), we get

$$(155) \qquad v = w - \frac{1}{\gamma}\Psi, \quad \Psi_z = 2\gamma_z w, \quad \Psi_{\bar{z}} = 2\gamma w_{\bar{z}} .$$

In this connection, for example, we consider the differential equation

$$(156) \qquad (\alpha+\bar{\alpha})^2 RSw - n(n+1)w = 0, \quad n \in \mathbb{N},$$

where $\alpha(z)$ denotes a holomorphic function in D with $(\alpha+\bar{\alpha})\alpha' \neq 0$ and

$$R = \frac{1}{\alpha'}\frac{\partial}{\partial z}, \quad S = \frac{1}{\overline{\alpha'}}\frac{\partial}{\partial \bar{z}} .$$

The solutions of (156) defined in D can be represented by using Theorem 9; it follows

$$(157) \qquad w = Hg + \overline{Hh},$$

where $g(z)$, $h(z) \in H(D)$ and

$$H = \sum_{k=0}^{n} \frac{A_k^n}{(\alpha+\bar{\alpha})^{n-k}} R^k, \quad A_k^n = \frac{(-1)^{n-k}(2n-k)!}{k!(n-k)!} .$$

If γ is a particular nonvanishing solution of (156) in D, the corresponding differential equation (153) runs

$$(158) \qquad (\alpha+\bar{\alpha})^2 RSv + \left[n(n+1) - \frac{2(\alpha+\bar{\alpha})^2 R\gamma S\gamma}{\gamma^2} \right] v = 0.$$

For instance, if we set $\alpha(z) = z$, we get

$$(159) \qquad (z+\bar{z})^2 v_{z\bar{z}} + \left[n(n+1) - 2\,\frac{(z+\bar{z})^2 \gamma_z \gamma_{\bar{z}}}{\gamma^2} \right] v = 0.$$

Because of the appearance of the first and second Beltrami operator applied to γ and v respectively this differential equation has certain invariance properties with regard to the automorphisms of the right half-plane. Therefore, the solutions of (159) are of interest in the theory of automorphic functions.

If we suppose that the particular solution γ of (156) has the representation

$$\gamma = H\varphi + \overline{H\psi}\ , \qquad \varphi(z), \psi(z) \in H(D),$$

and if we use

$$w = Hg \text{ with } g(z) \in H(D),$$

first, by (155) we get

$$\psi = \sum_{k=0}^{n-1} [p_k(\varphi) + \overline{q_k(\psi)}] \frac{A_k^n R^k g}{(\alpha+\bar{\alpha})^{n-k}} + r(z) \tag{160}$$

with

$$p_k(\varphi) = \sum_{s=0}^{n} \frac{2(n-k)A_s^n}{2n-k-s} \frac{R^s \varphi}{(\alpha+\bar{\alpha})^{n-s}}\ ,$$

$$q_k(\psi) = \sum_{s=0}^{n-1} B_s^k \frac{R^s \psi}{(\alpha+\bar{\alpha})^{n-s}}\ ,$$

$$B_s^k = 2(k-n) \sum_{\mu=0}^{n-s-1} \frac{(2n-s-k-1)!}{(n+\mu-k)!} A_{n-\mu}^n\ ,$$

where r(z) denotes, for the present, an arbitrary holomorphic function in D. In order to determine this function we substitute (160) into $\psi_z = 2w\gamma_z$ and obtain

$$v = Hg - \frac{1}{\gamma} K_{\varphi\psi} g$$

with

$$K_{\varphi\psi} g = \sum_{k=0}^{n-1} [p_k(\varphi) + \overline{q_k(\psi)}] \frac{A_k^n R^k g}{(\alpha+\bar{\alpha})^{n-k}} + 2 \int \alpha' R^{n+1} \varphi R^n g dz.$$

Because of

$$v = -u + \frac{1}{\gamma}(2\gamma u - \psi)$$

every solution of the form

$$v = u - \frac{1}{\gamma}\psi$$

may be represented also by

$$v = -u + \frac{1}{\gamma}\Phi .$$

Therefore, it is advantageous to determine the second part of the solution by (154). Considering Theorem 9,c we get the following general representation theorem for the solutions of (158) defined in D (cf. [19]).

Theorem 22

Let $\gamma = H\varphi + \overline{H\psi}$ be a nonvanishing particular solution of (156) defined in D.

a) For every solution v of the differential equation (158) defined in D there exist two functions $g(z), h(z) \in H(D)$, such that

$$v = Hg + \overline{Hh} - \frac{1}{\gamma}[K_{\varphi\psi}g + \overline{K_{\psi\varphi}h}]. \tag{161}$$

b) Conversely, for arbitrary functions $g(z), h(z) \in H(D)$ (161) represents a solution of (158) in D.

c) For every given solution v of (158) the quantities $R[\gamma^{-1}(Hg-\overline{Hh}]$ and $S[\gamma^{-1}(Hg-\overline{Hh})]$ are uniquely determined by

$$R[\gamma^{-1}(Hg-\overline{Hh})] = \gamma^{-2}R(\gamma v), \quad S[\gamma^{-1}(Hg-\overline{Hh})] = -\gamma^{-2}S(\gamma v).$$

In this case the generators g(z) and h(z) are not uniquely determined. We obtain the most general pair of generators $\tilde{g}(z)$ and $\tilde{h}(z)$ by

$$\tilde{g} = g + a\varphi + g_o, \quad \tilde{h} = h - \bar{a}\psi + h_o$$

with

$$g_o = \sum_{\mu=0}^{2n} a_\mu \alpha^\mu , \qquad h_o = \sum_{\mu=0}^{2n} (-1)^\mu \overline{a_\mu} \alpha^\mu ,$$

$$a, a_\mu \in \mathbb{C}, \qquad 2\gamma H g_o = K_{\varphi\psi} g_o + \overline{K_{\psi\varphi} h_o} .$$

d) If γ is a real-valued solution of (156) with

$$\gamma = H\varphi + \overline{H\varphi} ,$$

we get the real-valued solutions v of (158) defined in D by

$$v = Hf + \overline{Hf} - \frac{1}{\gamma}[K_{\varphi\varphi} f + \overline{K_{\varphi\varphi} f}], \qquad f(z) \in H(D).$$

These representations of the solutions v may be simplified considerably if the generators of γ reduce to polynomials in α of degree 2n. In this case we can attain that the terms $q_k(\psi)$ in $K_{\varphi\psi}$ vanish. Moreover, by repeated integration by parts the representation may be converted in a form free of integrals. It is also possible to get such a form if the generators of γ are not polynomials in α of degree 2n (cf.[19]). Presumably there exist further functions γ for which the solutions v of (158) can be represented by differential operators only. However, a general characterizing of all solutions γ of (156) with this property is lacking.

5) A generalized Darboux equation

G.Darboux treated in [40], L.IV, Ch.III, the hyperbolic differential equation

$$\Phi_{xy} = \frac{\mu(1-\mu)}{(x-y)^2}\,\Phi \tag{162}$$

which has been object of much study, in part due to its special mathematical properties and in part due to its appearance in many specific problems in classical physics (cf. e.g.[50,51,52,113]). The differential equation (162) which is often termed Euler-Poisson-Darboux equation has been generalized in different ways. On the one hand the corresponding differential equation with m independent variables has been treated in a number of papers (cf. e.g. [37,41,103,111]), on the other hand G.Darboux considered in [40], L.IV, Ch.IX, a differential equation in which the coefficient of Φ represents an essentially more general function. If we replace in (162)

$$x \text{ by } -x,\ \frac{1}{x}\ ,\ \text{and}\ -\frac{1}{x}$$

the quotient $\Phi^{-1}\Phi_{xy}$ equals

$$-\frac{\mu(1-\mu)}{(x+y)^2}\ ,\ -\frac{\mu(1-\mu)}{(1-xy)^2}\ ,\ \text{and}\ \frac{\mu(1-\mu)}{(1+xy)^2}$$

respectively. That gave rise to consider the hyperbolic differential equation

$$\frac{1}{\Phi}\,\Phi_{xy} = \frac{\mu(1-\mu)}{(x-y)^2} - \frac{\mu'(1-\mu')}{(x+y)^2} - \frac{\nu(1-\nu)}{(1-xy)^2} + \frac{\nu'(1-\nu')}{(1+xy)^2}\ , \tag{163}$$

$$\mu, \mu', \nu, \nu' \in \mathbb{R}.$$

Among other things G.Darboux could show that the differential equation (163) can be integrated for $\mu, \mu', \nu, \nu' \in \mathbb{Z}$. In this context it is sufficient to suppose

$$\mu, \mu', \nu, \nu' \in \mathbb{N}_0. \tag{164}$$

The case of negative integers can be reduced to (164).
If we formally replace

$$x \text{ by } z = x + iy$$

and

$$y \text{ by } \bar{z} = x - iy,$$

we get the differential equation

$$w_{z\bar{z}} + \left[\frac{-m(m+1)}{(z+\bar{z})^2} + \frac{n(n+1)}{(z-\bar{z})^2} - \frac{p(p+1)}{(1-z\bar{z})^2} + \frac{q(q+1)}{(1+z\bar{z})^2} \right] w = 0 \tag{165}$$

with

$$m,n,p,q \in \mathbb{N}_0 \; .$$

Using Theorem 15 and Theorem 16, we can get representations of solutions of (165) by differential operators. And here the generators are arbitrary solutions of a differential equation of the same type, where the coefficient of w has less summands than the coefficient in the differential equation in question. Such representations can be found in those cases in which the coefficients of w has up to three summands. In the case of four summands a corresponding simple representation is possible if two of the parameters are equal.
We demonstrate this procedure by the following example and refer the reader to [16] for further details.
First, we set

$$\gamma = \delta = z$$

and obtain by Theorem 15

$$K_1 = r + s, \quad K_1 B = 0.$$

If we use, for example, $B \equiv 0$, we get a solution of the differential equation

$$u_{z\bar{z}} - \frac{m(m+1)}{(z+\bar{z})^2} u = 0, \quad m \in \mathbb{N}_0, \tag{166}$$

defined in D by

$$u = \sum_{k=0}^{m} \frac{(-1)^{m-k}(2m-k)!}{k!(m-k)!} \frac{K_1^k h}{(z+\bar{z})^{m-k}},$$

where h denotes an arbitrary solution of the differential equation $h_{z\bar{z}} = 0$ in D.

Setting

$$\gamma = z \text{ and } \delta = -z,$$

it follows by Theorem 15

$$K_2 = r - s, \quad K_2 B = 0.$$

If we use

$$B = \frac{-m(m+1)}{(z+\bar{z})^2},$$

we obtain a solution of the differential equation

(167) $$v_{z\bar{z}} + \left[\frac{-m(m+1)}{(z+\bar{z})^2} + \frac{n(n+1)}{(z-\bar{z})^2}\right] v = 0, \quad m,n \in \mathbb{N}_0,$$

defined in D by

$$v = \sum_{j=0}^{n} \frac{(-1)^{n-j}(2n-j)!}{j!(n-j)!} \frac{K_2^j u}{(z-\bar{z})^{n-j}},$$

where u is an arbitrary solution of (166) in D.

Finally, setting

$$\gamma = \frac{1}{z} \text{ and } \delta = \varepsilon z, \quad \varepsilon = \pm 1,$$

it follows by Theorem 16

$$K_3 = zr + \bar{z}s, \quad K_3 B + 2B = 0.$$

If we now use

$$B = -\frac{m(m+1)}{(z+\bar{z})^2} + \frac{n(n+1)}{(z-\bar{z})^2},$$

we obtain a solution of the differential equation

$$w_{z\bar{z}} + \left[\frac{-m(m+1)}{(z+\bar{z})^2} + \frac{n(n+1)}{(z-\bar{z})^2} + \frac{\varepsilon p(p+1)}{(1+\varepsilon z\bar{z})^2}\right] w = 0$$

in D by

$$w = \sum_{s=0}^{p} \frac{(2p-s)!}{s!(p-s)!} \frac{(K_3-p)_s v}{(1+\varepsilon z\bar{z})^{p-s}},$$

where v is an arbitrary solution of (167) in D.

Theorem 23

Let D_ε, $\varepsilon = \pm 1$, be a simply connected domain of the complex plane with

$$(z^2-\bar{z}^2)(1+\varepsilon z\bar{z}) \neq 0.$$

Let h be an arbitrary solution of $h_{z\bar{z}} = 0$ in D_ε.
Then,

$$w = \sum_{s=0}^{p} \frac{(2p-s)!}{s!(p-s)!} \frac{(K_3-p)_s v}{(1+\varepsilon z\bar{z})^{p-s}}$$

with

$$v = \sum_{j=0}^{n} \frac{(-1)^{n-j}(2n-j)!}{j!(n-j)!} \frac{K_2^j u}{(z-\bar{z})^{n-j}}$$

and

$$u = \sum_{k=0}^{m} \frac{(-1)^{m-k}(2m-k)!}{k!(m-k)!} \frac{K_1^k h}{(z+\bar{z})^{m-k}}$$

represents a solution of the differential equation

(168) $$w_{z\bar{z}} + \left[\frac{-m(m+1)}{(z+\bar{z})^2} + \frac{n(n+1)}{(z-\bar{z})^2} + \frac{\varepsilon p(p+1)}{(1+\varepsilon z\bar{z})^2} \right] w = 0,$$

$$m,n,p \in \mathbb{N}_0, \quad \varepsilon = \pm 1,$$

in D_ε.

Other forms of the representation of solutions can be found by means of Theorem 17. As an example also here we consider the differential equation (168).
We use

$$\gamma = \delta = z, \quad \Psi_1(z+\bar{z}) = \frac{-m(m+1)}{(z+\bar{z})^2}, \quad m \in \mathbb{N}_0, \quad \Psi_2(\vartheta) \equiv 0$$

in Theorem 17 and get by $C_1 = 0$

$$\Phi = \frac{-2(m+1)}{z+\bar{z}}$$

as a particular solution of (142). By

$$\gamma = z, \ \delta = -z, \quad \Psi_1(z-\bar{z}) = \frac{-n(n+1)}{(z-\bar{z})^2}, \quad n \in \mathbb{N}_0,$$

$$\Psi_2(z+\bar{z}) = \frac{m(m+1)}{(z+\bar{z})^2}, \quad C_1 = 0,$$

it follows

$$\Phi = -\frac{2(n+1)}{z-\bar{z}}$$

as a particular solution of (142). Finally, by

$$\gamma=\delta=\log z, \quad \Psi_1(\log z\bar{z}) = \frac{\varepsilon p(p+1)z\bar{z}}{(1+\varepsilon z\bar{z})^2}, \quad p \in \mathbb{N}_0, \quad \varepsilon = \pm 1,$$

$$\Psi_2(\log \frac{z}{\bar{z}}) = \frac{-m(m+1)z\bar{z}}{(z+\bar{z})^2} + \frac{n(n+1)z\bar{z}}{(z-\bar{z})^2}, \quad m,n \in \mathbb{N}_0,$$

we get by $C_1 = -(p+1)^2$

$$\Phi = (p+1)\,\frac{1-\varepsilon z\bar{z}}{1+\varepsilon z\bar{z}}$$

as a particular solution of (142). By successive application of the differential operators (141), determined by the respective choice of γ and δ and the particular solution Φ, we obtain another representation for the solutions of (168).

Theorem 24

Let D_ε, $\varepsilon = \pm 1$, be a simply connected domain of the complex plane with

$$(z^2 - \bar{z}^2)(1+\varepsilon z\bar{z}) \neq 0.$$

Let h be a solution of $h_{z\bar{z}} = 0$ in D_ε. The operators $K_{1,m}$, $K_{2,n}$, and $K_{3,p}$ are defined by

$$K_{1,m} = r + s - \frac{2m}{z+\bar{z}}\,, \tag{169}$$

$$K_{2,n} = r - s - \frac{2n}{z-\bar{z}}\,, \tag{170}$$

$$K_{3,p} = zr + \bar{z}s + p\,\frac{1-\varepsilon z\bar{z}}{1+\varepsilon z\bar{z}}\,. \tag{171}$$

Then

$$w = K_{3,p} \cdots K_{3,1} v \tag{172}$$

with

$$v = K_{2,n} \cdots K_{2,1} u \tag{173}$$

and

$$u = K_{1,m} \cdots K_{1,1} h$$

represents a solution of (168)

$$w_{z\bar{z}} + \left[\frac{-m(m+1)}{(z+\bar{z})^2} + \frac{n(n+1)}{(z-\bar{z})^2} + \frac{\varepsilon p(p+1)}{(1+\varepsilon z\bar{z})^2} \right] w = 0$$

in D_ε. Here, in case $p = 0$, $n = 0$, and $m = 0$ respectively the operators

$$K_{3,p} \dots K_{3,1}, \quad K_{2,n} \dots K_{2,1}, \quad K_{1,m} \dots K_{1,1},$$

are to replace by the identity operator.

6) The differential equation $\omega^2 w_{z\bar{z}} + C\varphi'\overline{\psi'}w = 0$, $C \in \mathbb{C}$

In the following let the functions $\varphi(z)$ and $\psi(z)$ satisfy again the conditions (20); moreover, we use again

$$\omega = \varphi + \bar{\psi}, \quad R = \frac{1}{\varphi'}\frac{\partial}{\partial z}, \quad \text{and} \quad S = \frac{1}{\overline{\psi'}}\frac{\partial}{\partial \bar{z}}.$$

Then, by

(175) $$\omega^2 RSw + Cw = 0, \quad C \in \mathbb{C},$$

we have a differential equation which has been treated in a number of papers (cf. e.g. [3,28,46,86,95]). However, in the case

(176) $$C \neq -n(n+1), \quad n \in \mathbb{N}_0,$$

general representation theorems for the solutions are lacking.
In the following we give a procedure which allows, in a simple way, to determine explicitly a class of particular solutions of (175) with C according to (176). Here, homogeneous polynomials in φ and $\bar{\psi}$ of degree $m \in \mathbb{N}_0$ appear for which we can get certain functional-differential-relations.
First, we obtain by

(177) $$\begin{cases} w = C_1\omega^{\lambda_1} + C_2\omega^{\lambda_2}, & C_1, C_2 \in C, \\ \lambda_{1,2} = \frac{1}{2}[1 \pm \sqrt{1-4C}], & \lambda_{1,2} \notin \mathbb{Z}, \end{cases}$$

all solutions of (175) which depend on ω only. If we set

$$w = v\omega^{\lambda_k}, \quad k = 1,2,$$

v satisfies the differential equation

(178) $$\omega RSv + \lambda_k(R+S)v = 0.$$

In order to determine particular solutions of this differential equation we require that v satisfies, in addition, the differential equation

$$\varphi Rv + \bar{\psi} Sv = mv,$$

that means that v denotes a homogeneous function in φ and $\bar{\psi}$ of degree m, $m \in \mathbb{R}$. Then,

$$v = \varphi^m X\left(\frac{\bar{\psi}}{\varphi}\right)$$

and it follows that the function

$$Y(\xi) = X\left(\frac{\bar{\psi}}{\varphi}\right) \quad \text{with} \quad \varphi\xi = -\bar{\psi}$$

satisfies the hypergeometric differential equation

$$\xi(\xi-1)Y'' + [(a+b+1)\xi - c]Y' + abY = 0 \tag{179}$$

with

$$a = -m, \quad b = \lambda_k, \quad c = 1-m-\lambda_k.$$

Therefore, the homogeneous function v reduces to a homogeneous polynomial if, and only if, $m \in \mathbb{N}_0$. Considering the representations of solutions of the hypergeometric differential equation (cf. e.g.[72]), we get the following

Theorem 25

If $w = v\omega^{\lambda_k}$, $k = 1,2$, with

$$v = \varphi^m X\left(\frac{\bar{\psi}}{\varphi}\right), \quad m \in \mathbb{R},$$

is a solution of the differential equation (175), the function v reduces to a homogeneous polynomial in φ and $\bar{\psi}$ if, and only if, $m \in \mathbb{N}_0$. In this case we have:

$$v = p_m(\lambda_k) = \sum_{s=0}^{m} a_s(m,\lambda_k)\varphi^s \bar{\psi}^{m-s} \tag{180}$$

with

$$(181) \qquad a_s(m,\lambda_k) = \binom{m}{s} \frac{(1+s-\lambda_k-m)_{m-s}}{(s+\lambda_k)_{m-s}}, \qquad s = 0,1, \ldots, m.$$

For the coefficients defined by (181) we get

$$(182) \qquad a_s(m,\lambda_k) = (-1)^m a_{m-s}(m,\lambda_k),$$

as can easily be verified. On account of $C = -(\lambda_k-1)\lambda_k$ we denote by $F_{\lambda_k-1}(D)$ the solutions of (175) in D, analogous to the notation used in Theorem 19.

Moreover, we set

$$L = R - S, \qquad L^* = R + S,$$

$$L_\nu = L^* - \frac{\nu}{\omega}, \qquad \nu \in \mathbb{R}.$$

Then, it follows (cf. Theorem 19) for a solution $w \in F_{\lambda_k-1}(D)$:

$$(183) \qquad Lw \in F_{\lambda_k-1}(D),$$

$$(184) \qquad L_{\lambda_k} w \in F_{\lambda_k}(D),$$

$$(185) \qquad L_{1-\lambda_k} w \in F_{\lambda_k-2}(D).$$

Using for w in (183) - (185) the solutions

$$w = \omega^{\lambda_k} p_m(\lambda_k), \qquad k = 1,2, \qquad m \in \mathbb{N}_0,$$

of Theorem 25, we get again solutions of this class. Moreover, we obtain certain functional-differential-relations for the homogeneous polynomials $p_m(\lambda_k)$.

If $m \in \mathbb{N}$, first it follows by (183)

$$L(\omega^{\lambda_k} p_m(\lambda_k)) = \omega^{\lambda_k} Lp_m(\lambda_k) \in F_{\lambda_k - 1}.$$

Since $Lp_m(\lambda_k)$ represents a homogeneous polynomial of degree $m-1 \in \mathbb{N}_o$, by Theorem 25 there exists a constant $c_1 \in \mathbb{C}$, such that

$$Lp_m(\lambda_k) = c_1 p_{m-1}(\lambda_k). \tag{186}$$

We get this constant immediately by comparison of the coefficients of φ^{m-1}; it follows

$$c_1 = ma_m(m,\lambda_k) - a_{m-1}(m,\lambda_k) = \frac{m(m+2\lambda_k-1)}{m+\lambda_k-1}. \tag{187}$$

Using (184) with $m \geq 2$, it follows

$$L_{\lambda_k}(\omega^{\lambda_k} p_m(\lambda_k)) = \omega^{\lambda_k+1} \frac{L^* p_m(\lambda_k)}{\omega} \in F_{\lambda_k}.$$

Since $v = \frac{1}{\omega} L^* p_m(\lambda_k)$ is a solution of the differential equation

$$\varphi R v + \bar{\varphi} S v = (m-2) v$$

and, therefore, a homogeneous polynomial of degree $m-2 \in \mathbb{N}_o$, by Theorem 25 it follows that there exists a constant $c_2 \in \mathbb{C}$, such that

$$L^* p_m(\lambda_k) = c_2 \omega p_{m-2}(\lambda_k+1). \tag{188}$$

A comparison of the coefficients of φ^{m-1} yields here

$$c_2 = ma_m(m,\lambda_k) + a_{m-1}(m,\lambda_k) = \frac{m(m-1)}{m+\lambda_k-1}. \tag{189}$$

Finally, we get by (185) for $m \in \mathbb{N}$

$$L_{1-\lambda_k}(\omega^{\lambda_k} p_m(\lambda_k)) = \omega^{\lambda_k-1} q_m(\lambda_k) \in F_{\lambda_k-2}$$

with

$$q_m(\lambda_k) = [2(2\lambda_k-1) + \omega L^*]p_m(\lambda_k),$$

where $q_m(\lambda_k)$ denotes a homogeneous polynomial in φ and $\bar{\psi}$ of degree m. Hence, there exists a constant $c_3 \in \mathbb{C}$, such that

$$(190) \qquad q_m(\lambda_k) = c_3 p_m(\lambda_k-1),$$

and it follows by comparison of the coefficients of φ^m

$$(191) \qquad c_3 = a_m(m,\lambda_k)[m+2(2\lambda_k-1)] + a_{m-1}(m,\lambda_k) = 2(2\lambda_k-1) + \frac{m(m-1)}{m+\lambda_k-1} .$$

By suitable combination of (186), (188), and (190) further relations result, and summarizing we get the following

Theorem 26

If $\omega = \varphi + \bar{\psi}$ and

$$p_m(\lambda_k) = \sum_{s=0}^{m} a_s(m,\lambda_k)\varphi^s \bar{\psi}^{m-s}, \qquad m \in \mathbb{N}_o, \qquad k = 1,2,$$

with $a_s(m,\lambda_k)$ and λ_k according to (181) and (177) respectively, then,

1) $a_s(m,\lambda_k) = (-1)^m a_{m-s}(m,\lambda_k)$,

2) $(m+\lambda_k-1)(R-S)p_m(\lambda_k) = m(m+2\lambda_k-1)p_{m-1}(\lambda_k), \quad m \geq 1$,

3) $(m+\lambda_k-1)(R+S)p_m(\lambda_k) = m(m-1)\omega p_{m-2}(\lambda_k+1), \quad m \geq 2$,

4) $(m+\lambda_k-1)[2(2\lambda_k-1) + \omega(R+S)]p_m(\lambda_k) =$

$= [m(m-1) + 2(2\lambda_k-1)(m+\lambda_k-1)]p_m(\lambda_k-1), \quad m \geq 1$

5) $2(m+\lambda_k-1)Rp_m(\lambda_k) = m(m-1)\omega p_{m-2}(\lambda_k+1)+m(m+2\lambda_k-1)p_{m-1}(\lambda_k), \; m \geq 2$,

6) $2(m+\lambda_k-1)Sp_m(\lambda_k) = m(m-1)\omega p_{m-2}(\lambda_k+1)-m(m+2\lambda_k-1)p_{m-1}(\lambda_k), \; m \geq 2$,

7) $2(m+\lambda_k-1)(2\lambda_k-1+\omega R)p_m(\lambda_k) =$

$=[m(m-1)+2(2\lambda_k-1)(m+\lambda_k-1)]p_m(\lambda_k-1)+m(m+2\lambda_k-1)\omega p_{m-1}(\lambda_k), \; m \geq 1$,

8) $2(m+\lambda_k-1)(2\lambda_k-1+\omega S)p_m(\lambda_k) =$

$$=[m(m-1)+2(2\lambda_k-1)(m+\lambda_k-1)]p_m(\lambda_k-1)-m(m+2\lambda_k-1)\omega p_{m-1}(\lambda_k), \quad m \geq 1,$$

9) $$2(m+\lambda_k-1)(2\lambda_k-1)p_m(\lambda_k)+m(m-1)\omega^2 p_{m-2}(\lambda_k+1) =$$

$$=[m(m-1)+2(2\lambda_k-1)(m+\lambda_k-1)]p_m(\lambda_k-1), \quad m \geq 2.$$

Proceeding from the solutions

$$w = \omega^{\lambda_k} p_m(\lambda_k), \quad k = 1,2 \quad m \in \mathbb{N}_0, \tag{192}$$

of the differential equation (175) (cf. Theorem 25), we may get solutions of further differential equations if we apply Theorem 17. We set, for example,

$$\gamma = \varphi, \quad \delta = -\psi, \quad \Psi_1 = \frac{-n(n+1)}{\eta^2}, \quad \eta = \varphi-\bar{\varphi}, \quad \Psi_2 = -C.$$

Then, from a solution w of the differential equation

$$RSw + \left[\frac{C}{\omega^2} + \frac{n(n+1)}{\eta^2}\right] w = 0$$

we obtain by

$$v = (R-S+\Phi)w$$

a solution of the differential equation

$$RSv + \left[\frac{C}{\omega^2} + \frac{n(n+1)}{\eta^2} + \Phi'(\eta)\right] v = 0,$$

if $\Phi(\eta)$ is a solution of the Riccati equation

$$2\Phi' = \Phi^2 - \frac{4n(n+1)}{\eta^2} + C_1, \quad C_1 \in \mathbb{C}. \tag{193}$$

Setting $C_1 = 0$, by

$$\Phi = -\frac{2(n+1)}{\eta}$$

we get a particular solution of (193). Thus, by successive application of the operators

$$R-S-\frac{2s}{\eta}, \quad s = 1,2, \ldots, n,$$

from the solutions (192) we obtain solutions of the differential equation

$$RSv + \left[\frac{C}{\omega^2} + \frac{n(n+1)}{\eta^2}\right] v = 0.$$

Similarly by the operators

$$R - \frac{\nu}{\omega} \quad \text{and} \quad \omega S - \mu , \quad \nu,\mu \in \mathbb{C},$$

from the functions (192) we get solutions of differential equations of the form

$$\omega^2 RSv + \omega C_1 Sv + C_2 v = 0, \quad C_1, C_2 \in \mathbb{C}.$$

7) Differential operators for a class of elliptic differential equations of even order

By application of Theorem 11 we get the possibility to obtain general representation theorems for the solutions of a class of differential equations of even order which are defined in simply connected domains or in the neighbourhood of isolated singularities. These results were derived in [12] and [30].

We consider the differential equation

$$(194) \qquad T_1 T_2 \ldots T_m w = 0$$

with

$$(195) \qquad T_k = \omega^2 RS - n_k(n_k+1), \quad n \in \mathbb{N}_0, \quad n_k \neq n_j.$$

Here, we use again

$$\omega = \varphi + \overline{\psi}, \quad R = \frac{1}{\varphi'} \frac{\partial}{\partial z}, \quad S = \frac{1}{\overline{\psi'}} \frac{\partial}{\partial \overline{z}}$$

and suppose that the functions $\varphi(z)$ and $\psi(z)$ satisfy the conditions (20) in a simply connected domain D.

For the solutions of the differential equation (194) I.N. Vekua derived a representation by means of integral operators (cf. [102], Ch.V). Indepently in [12] and [30] representations were proved which are free of integrals; here, the generators are mapped onto solutions of (194) by differential operators. In general the parameter n_k in (195) may be an arbitrary element of the set $\mathbb{N}_0$. However, if the differential equation has the special form

$$(196) \qquad T_0 T_1 \ldots T_n w = 0$$

with

$$(197) \qquad T_s = (\varphi+\overline{\psi})^2 RS - s(s+1), \quad s = 0, 1, \ldots, n,$$

$$\varphi(z), \psi(z) \in H(D), \quad (\varphi+\overline{\psi})\varphi'\psi' \neq 0 \text{ in } D,$$

it is possible to simplify the results essentially (cf. [30]). These representations are especially suitable, for instance, for the research of a certain subset of the solutions of (196) which may be termed as generalized holomorphic functions. In contrast the representation derived in [12] is advantageous, for example, to get assertions about the real and imaginary part of the solutions of (194) in case of real-valued coefficients

$$F = \frac{\omega^2}{\varphi'\overline{\psi'}} .$$

For instance, necessary and sufficient conditions can be derived for the generators of the solutions of (194) if these solutions are equal in their real and imaginary parts respectively. Moreover, necessary and sufficient conditions can be determined for the generators of real-valued solutions of (194) for the case that these solutions arise as real respect. imaginary part of a complex-valued solution of (194).

In the following we sketch out the procedure applied in [12] and [30] and summarize some of the derived results.

Proceeding from (194), we set

$$W_1 = T_2 \ldots T_m w.$$

Then, W_1 satisfies the differential equation

$$T_1 W_1 = 0$$

whose solutions in D may be represented by Theorem 9 in the form

$$W_1 = H_{n_1} f_1 + H^*_{n_1} \overline{f^*_1} \tag{198}$$

with $f_1 \in M_{2n_1}(\varphi, D)$, $f^*_1 \in M_{2n_1}(\psi, D)$. By

$$W_2 = T_3 \ldots T_m w$$

we get for W_2 the inhomogeneous differential equation

$$T_2 W_2 = W_1$$

with W_1 according to (198). Applying Theorem 11, the solutions of this differential equation may be represented by

$$W_2 = H_{n_2} f_2 + H^*_{n_2} \overline{f^*_2} + \frac{1}{n_1(n_1+1)-n_2(n_2+1)} W_1$$

with $f_2 \in M_{2n_2}(\varphi,D)$, $f^*_2 \in M_{2n_2}(\psi,D)$. Generally, we set

$$W_s = T_{s+1}T_{s+2} \cdots T_m w,$$

and obtain the inhomogeneous differential equations

$$T_s W_s = W_{s-1} ,$$

whose solutions can be represented in each case by means of Theorem 11. Thus, we obtain the following

Theorem 27

a) For every solution of the differential equation (194)

$$T_1 T_2 \cdots T_n w = 0$$

with

$$T_k = \omega^2 RS - n_k(n_{k+1}), \quad n_k \in \mathbb{N}_o, \quad n_k \neq n_j ,$$

defined in D, there exist 2m functions

$$(199) \qquad g_k(z) \in M_{2n_k}(\varphi,D), \quad h_k(z) \in M_{2n_k}(\psi,D), \quad k = 1, \ldots, m,$$

such that

$$(200) \qquad w = \sum_{k=1}^{m} w_k \quad \text{with} \quad w_k = H_{n_k} g_k + H^*_{n_k} \overline{h_k} .$$

b) Conversely, (200) represents a solution of (194) in D if the generators $g_k(z)$ and $h_k(z)$ satisfy the condition (199).

c) For every given solution w of (194) the functions

$$w_k, \quad R^{2n_k+1} g_k, \quad \text{and} \quad S^{2n_k+1}\overline{h_k}$$

are uniquely determined by

$$w_k = \lambda_{(k)}^{-1} T_{(k)} w,$$

$$R^{2n_k+1} g_k(z) = \frac{P^{n_k+1}(T_{(k)}w)}{\lambda_{(k)}\omega^{2n_k+2}}, \tag{201}$$

$$S^{2n_k+1}\overline{h_k(z)} = \frac{Q^{n_k+1}(T_{(k)}w)}{\lambda_{(k)}\omega^{2n_k+2}} \tag{202}$$

with

$$\lambda_{(k)} = \prod_{\substack{s=1\\ s\neq k}}^{m} [n_k(n_k+1)-n_s(n_s+1)],$$

$$T_{(k)} = T_1 \dots T_{k-1}T_{k+1} \dots T_m ,$$

$$P = \omega^2 R, \quad Q = \omega^2 S.$$

For every given solution w the generators $g_k(z)$ and $h_k(z)$ are only determined up to polynomials in φ respectively ψ of degree $2n_k$. We obtain the most general generators $\tilde{g}_k(z)$ and $\tilde{h}_k(z)$ by

$$\tilde{g}_k(z) = g_k(z) + p_{k1}(\varphi), \quad \tilde{h}_k(z) = h_k(z) + p_{k2}(\psi)$$

with

$$p_{k1}(\varphi) = \sum_{\mu=0}^{2n_k} c_{k\mu}\varphi^{\mu}, \quad p_{k2}(\psi) = \sum_{\mu=0}^{2n_k} (-1)^{\mu+1}\overline{c_{k\mu}}\psi^{\mu}, \quad c_{k\mu} \in \mathbb{C}.$$

d) For every solution w of (194) in D which can be represented in the form

$$w = \sum_{k=1}^{m} H_{n_k} g_k \quad \text{respectively} \quad w = \sum_{k=1}^{m} H^{*}_{n_k} \overline{h_k}$$

the generators are uniquely determined by

$$g_k(z) = \frac{Q^{n_k}(T_{(k)}w)}{\lambda_{(k)}(2n_k)!}, \quad h_k(z) = \frac{P^{n_k}(T_{(k)}w)}{\lambda_{(k)}(2n_k)!}.$$

<u>Corollary</u>

Because of (201) and (202) for every solution w of (194) defined in a (not necessarily simply connected) domain D the functions

$$R^{2n_k+1} g_k(z) \quad \text{and} \quad \overline{S^{2n_k+1}\overline{h_k(z)}}, \quad k = 1, \ldots, m,$$

are uniquely determined in each point of D and represent globally unique holomorphic functions in D.

We get a corresponding representation theorem if the operators T_k in (195) have the special form

$$T_k = \frac{(1+\varepsilon f\overline{f})^2}{-\varepsilon f'\overline{f'}} \frac{\partial^2}{\partial z \partial \overline{z}} - n_k(n_k+1), \quad n_k \in \mathbb{N}, \quad n_k \neq n_j,$$

where the function f(z) satisfies the conditions (22). In this case the number of generators reduces to the half (cf.[12], Theorem 2). Proceeding from the above Corollary a general representation of the solutions of (194) in the neighbourhood of isolated singularities may be derived corresponding to the assertion in Theorem 7 (cf.[12], Theorem 3).

If the differential equation (194) has the special form (196), where the functions $\varphi(z)$ and $\psi(z)$ satisfy the conditions (197), we can get an essentially simpler representation of the solutions. Proceeding from (194) we obtain the differential equation (196) by

$$n_k = k-1, \quad k = 1,2, \ldots, m.$$

Arranging the corresponding representation of the solutions (200) with respect to powers of $\omega = \varphi + \bar{\psi}$, we obtain

$$w = \sum_{s=0}^{n} \frac{g_s^*(z) + \overline{h_s^*(z)}}{\omega^{n-s}}$$

with

$$g_s^*(z) = \sum_{k=n-s}^{n} a_{nk} R^{k+n-s} g_{k+1} ,$$

$$h_s^*(z) = \sum_{k=n-s}^{n} a_{nk} S^{k+n-s} \overline{h_{k+1}} ,$$

where the coefficients a_{nk} denote certain nonvanishing real numbers. By a detailed investigation (cf. [30]) we get the following assertions if we use again g_s and h_s instead of g_s^* and h_s^*.

Theorem 28

a) For every solution of the differential equation (196)

$$T_0 T_1 \ldots T_n w = 0$$

with

$$T_s = (\varphi+\bar{\psi})^2 RS - s(s+1), \quad s = 0,1, \ldots, n,$$

$$\varphi(z), \psi(z) \in H(D), \quad (\varphi+\bar{\psi})^2 \varphi'\psi' \neq 0 \text{ in } D,$$

defined in D, there exist 2n+2 generators

$$g_s(z), h_s(z) \in H(D), \quad s = 0,1, \ldots, n,$$

such that

$$w = \sum_{s=0}^{n} \frac{g_s(z) + \overline{h_s(z)}}{\omega^{n-s}}, \qquad \omega = \varphi + \overline{\varphi}. \tag{203}$$

b) Conversely, (203) represents a solution of (196) in D if the generators $g_s(z)$ and $h_s(z)$ are holomorphic in D.

c) For every given solution w of (196) the functions $R^{2n+1-s}g_s$ and $S^{2n+1-s}\overline{h_s}$, $s = 0,1, \ldots, n$, are uniquely determined. In this case the generators $g_s(z)$ and $h_s(z)$ are only determined up to polynomials $p_s(\varphi)$ and $q_s(\psi)$ of degree 2n-s. We obtain the most general generators $\tilde{g}_s(z)$ and $\tilde{h}_s(z)$ by

$$\tilde{g}_s(z) = g_s(z) + p_s(\varphi), \quad \tilde{h}_s(z) = h_s(z) + q_s(\psi)$$

with

$$\sum_{s=0}^{n} [p_s(\varphi) + \overline{q_s(\psi)}]\omega^s \equiv 0.$$

d) For every given solution w of (196) which can be represented only by the generators $g_s(z)$ respectively $h_s(z)$ these functions are uniquely determined by

$$g_s(z) = \sum_{\mu=0}^{n-s} \frac{(-1)^\mu \omega^\mu}{s!\mu!} S^{\mu+s}(\omega^n w) \tag{204}$$

and

$$h_s(z) = \sum_{\mu=0}^{n-s} \frac{(-1)^\mu \omega^\mu}{s!\mu!} R^{\mu+s}(\omega^n \overline{w}). \tag{205}$$

Those solutions of (196) which may be represented only by the generators $g_s(z)$, $s = 0,1, \ldots, n$, represent, in a certain sense, generalized holomorphic functions. If we set n = 0 in

$$w = \sum_{s=0}^{n} \frac{g_s(z)}{\omega^{n-s}}, \quad g_s(z) \in H(D),$$

we get the holomorphic function $g_o(z)$. By $n \in \mathbb{N}_o$, in general, we have the solutions of the differential equation

$$Q^{n+1}w = 0, \quad n \in \mathbb{N}_o, \quad Q = \omega^2 S, \tag{206}$$

defined in D. Thus, (206) represents a generalized Cauchy-Riemann equation (cf. [6]) which reduces to the classical Cauchy-Riemann equation $w_{\bar{z}} = 0$ by $n = 0$. In the following we summarize some properties of the solution of (206) and refer the reader to [30] for further details. For a fixed $n \in \mathbb{N}_o$ the set of the solutions of (206) defined in D forms an (n+1)-dimensional vector space over the holomorphic functions with the basis

$$1, \frac{1}{\omega}, \frac{1}{\omega^2}, \ldots, \frac{1}{\omega^n}.$$

If we imploy the usual addition and multiplication of complex-valued functions, we obtain by

$$w = \sum_{s=0}^{n} \frac{g_s(z)}{\omega^{n-s}}, \quad n \in \mathbb{N}_o, \tag{207}$$

an associative algebra over $\mathbb{C}$. If we use in (207)

$$g_s(z) = a_s(z-z_o)^k, \quad s = 0,1, \ldots, n, \quad k \in \mathbb{Z},$$

by

$$\alpha = (a_o, a_1, \ldots, a_n)$$

we get a solution

$$Z_n^{(k)}(\alpha, z, z_o) = (z-z_o)^k \sum_{s=0}^{n} \frac{a_s}{\omega^{n-s}}$$

which may be termed formal power. By $n = 0$ this formal power reduces to the classical power

$$Z_o^{(k)}(\alpha, z, z_o) = a_o(z-z_o)^k.$$

Generally we get the following assertion.

Theorem 29

Every solution of (206) defined in a neighbourhood

$$U_\varepsilon(z_o) = \{ z \mid |z-z_o| < \varepsilon \}$$

permits a unique expansion of the form

$$w = \sum_{k=0}^{\infty} Z_n^{(k)}(\alpha_k, z, z_o), \quad \alpha_k = (a_{ok}, a_{1k}, \dots, a_{nk})$$

which converges in $U_\varepsilon(z_o)$.

If we consider that the holomorphic generators can be continued in D in only one way by the circle-chain method, by Theorem 29 immediately it follows the following identity theorem.

Theorem 30

If two solutions w_1 and w_2 of (206) defined in D coincide in $U_\varepsilon(z_o) \subset D$, it follows that w_1 and w_2 coincide throughout D.

As is well known in the classical function theory it is sufficient to suppose that the two functions coincide only in an infinite set of points having a limit point in D. Here, this supposition is not sufficient since the zeros are not necessarily isolated, as can easily be shown by simple examples. However, we get a generalized Laurent series analogous to the above formal power series.

Theorem 31

Let w be a solution of (206) in $\dot{U}_\varepsilon(z_o)$ with an isolated singularity at z_o. Then, w can be expanded into a formal Laurent series

$$w = \sum_{-\infty}^{\infty} Z_n^{(k)}(\alpha_k, z, z_o)$$

which converges in $\dot{U}_\varepsilon(z_o)$.

8) Differential equations in several independent complex variables

So far we have considered elliptic differential equations whose solutions can be generated by means of differential operators acting on functions which are defined in simply connected domains of the complex plane. Corresponding results can be derived also for differential equations in several independent complex variables. Here, the generators are functions which are defined in polydomains of the space $\mathbb{C}^2$ respectively $\mathbb{C}^m$. In the following this shall be demonstrated by some examples.

If we formally replace in (92) z by z_1 and $\bar{z}$ by z_2, we obtain the differential equation

$$(208) \qquad (1+z_1z_2)^2 w_{z_1z_2} + n(n+1)w = 0, \quad n \in \mathbb{N},$$ [9]

with

$$z_k = x_k + iy_k\,, \quad (\quad)_{z_k} = \frac{\partial}{\partial z_k} = \frac{1}{2}\left(\frac{\partial}{\partial x_k} - i\,\frac{\partial}{\partial y_k}\right), \quad k = 1,2,$$

where $w = w(z_1,z_2)$ denotes a holomorphic function of the two complex variables z_1 and z_2.

Setting $w = u + iv$ and

$$\varphi = \mathrm{Re}\{(1+z_1z_2)^2\} = (1+x_1x_2-y_1y_2)^2 - (x_1y_2+x_2y_1),$$

$$\psi = \mathrm{Im}\{(1+z_1z_2)^2\} = 2(1+x_1x_2-y_1y_2)(x_1y_2+x_2y_1),$$

the differential equation (208) is equivalent to the ultra-hyperbolic system of real differential equations

$$\varphi(u_{x_1x_2}-u_{y_1y_2}+v_{x_1y_2}+v_{x_2y_1})+\psi(u_{x_1y_2}+u_{x_2y_1}-v_{x_1x_2}+v_{y_1y_2})+4n(n+1)u = 0$$

$$\psi(u_{x_1x_2}-u_{y_1y_2}+v_{x_1y_2}+v_{x_2y_1})-\varphi(u_{x_1y_2}+u_{x_2y_1}-v_{x_1x_2}+v_{y_1y_2})+4n(n+1)v = 0$$

---

[9] Here, we only use $\varepsilon = 1$ since the case $\varepsilon = -1$ can be reduced to the case $\varepsilon = 1$ by a simple coordinate transformation.

defined in the space $\mathbb{R}^4$.

We denote by S the analytic surface of the space $\mathbb{C}^2$ which is given by

$$1 + z_1 z_2 = 0.$$

The space $\mathbb{C}^2$ is completed by means of the group

$$G: \zeta_k = \frac{a_k z_k + b_k}{c_k z_k + d_k}, \quad a_k d_k - b_k c_k \neq 0, \quad k = 1,2. \; ^{10)}$$

We denote by G* the following subgroup of G:

$$G^*: \begin{cases} \zeta_1 = \dfrac{a z_1 + b}{c z_1 + d} \\ \\ \zeta_2 = \dfrac{d z_2 - c}{-b z_2 + a} \end{cases} \qquad ad - bc \neq 0.$$

The differential equation (208) as well as the analytic surface S is invariant under all transformations of the group G*. Apart from the exchange of the variables z_1 and z_2, G* is the greates subgroup of G with this property. Moreover, G* contains, for its part, the two important subgroups

$$G_\varepsilon^*: \begin{cases} \zeta_1 = \eta \dfrac{z_1 - a}{1 + \varepsilon \bar{a} z_1} \\ \\ \zeta_2 = \bar{\eta} \dfrac{z_2 - \varepsilon \bar{a}}{1 + a z_2} \end{cases} \qquad \begin{array}{l} |\eta| = 1 \\ \varepsilon = +1: a \in \mathbb{C} \cup \{\infty\}, \\ \varepsilon = -1: |a| < 1. \end{array}$$

In the case of $\varepsilon = +1$ we have rotations of the Riemann number sphere in each complex plane; in the case of $\varepsilon = -1$ each transformation represents an automorphism of the unit disk in each complex plane. The open unit bicylinder, in particular, which is free of points of the surface S is mapped onto itself by each transformation of the group G_{-1}^*.

---

10) After Osgood the space $\mathbb{C}^2$ completed in this manner is termed the space of function theory (cf.[31]).

We denote by

$$D = D^{(1)} \times D^{(2)}$$

a polydomain of the space $\mathbb{C}^2$ which is the Cartesian product of the finite and simply connected domains $D^{(k)}$, $k = 1,2$, of the z_k-planes and which satisfies the condition

$$D \cap S = \emptyset .$$

Without loss of generality we can confine $D^{(k)}$ to finite domains. If such a domain contains the point at infinity, we can reduce it to the case treated here by a suitable transformation of the group G^*.

Formally in the same manner (cf. [3] and [5]) as in the case of the differential equation (92) we get the following general representation theorem.

Theorem 32

Let $D^{(k)}$, $k = 1,2$, be a finite and simply connected domain of the z_k-plane and

$$D = D^{(1)} \times D^{(2)} \text{ with } D \cap S = \emptyset.$$

a) For every holomorphic solution of the differential equation (208)

$$(1+z_1z_2)^2 w_{z_1z_2} + n(n+1)w = 0, \qquad n \in \mathbb{N},$$

defined in the polydomain D, there exist two functions $g_k(z_k)$ defined in $D^{(k)}$, $k = 1,2$, such that

$$w = E_1g_1 + E_2g_2 \tag{209}$$

with

$$E_k = \sum_{s=0}^{n} \frac{(-1)^{n-s}(2n-s)!}{s!(n-s)!} \left[\frac{z_j}{1+z_1z_2}\right]^{n-s} \frac{d^s}{dz_k^{\,s}}, \quad k = 1,2, \quad j \neq k.$$

b) Conversely, for each pair of functions $g_k(z_k)$, $k = 1,2$, holomorphic in $D^{(k)}$ (209) represents a holomorphic solution of (208) in the polydomain D.

c) For every given holomorphic solution $w = w(z_1,z_2)$ of (208) the derivatives of the generators of order 2n+1 are uniquely determined by

$$g_k^{(2n+1)}(z_k) = \frac{D_k^{n+1} w}{(1+z_1z_2)^{2n+2}}, \quad D_k = (1+z_1z_2)^2 \frac{\partial}{\partial z_k}, \quad k = 1,2.$$

In this case the generators $g_k(z_k)$ are only determined up to a polynomial $p(z_1)$ of degree 2n. We obtain the most general pair of generators by

$$\tilde{g}_1(z_1) = g_1(z_1) + p(z_1),$$

$$\tilde{g}_2(z_2) = g_2(z_2) - (-1)^n z_2^{2n} p\left(\frac{-1}{z_2}\right).$$

d) For every holomorphic solution $w = E_k g_k$, $k = 1,2$, defined in D the generator $g_k(z_k)$ is uniquely determined by

$$g_k(z_k) = \frac{(-1)^n}{(2n)!} D_j^n w, \quad k,j = 1,2, \quad j \neq k .$$

For the single-valued solutions of (208) we get the following assertion (cf.[5], Theorem 3).

Theorem 33

Let $D^{(k)}$, $k = 1,2$, be a finite (not necessarily simply connected) domain of the z_k-plane and

$$D = D^{(1)} \times D^{(2)} \text{ with } D \cap S = \emptyset .$$

Then, for every single-valued solution w of (208) defined in D there exist two single-valued holomorphic functions $g_k(z_k)$ defined in $D^{(k)}$, such that

$$w = E_1 g_1 + E_2 g_2.$$

However, this assertion is not valid for arbitrary domains of the space $\mathbb{C}^2$. For example, there exist single-valued solutions $w=E_1g_1+E_2g_2$ of (208) defined in circular regions (cf. e.g.[31]) of which the generators $g_k(z_k)$ are not single-valued. Corresponding examples are treated in [27]. And that a disk region K is constructed and it is shown that the function $w = E_1g_1 + E_2g_2$ defined in K and generated by the multi-valued functions

$$g_1(z_1) = \frac{1}{2\pi i}\, p(z_1)\log z_1$$

and

$$g_2(z_2) = \frac{(-1)^{n+1}}{2\pi i}\, z_2^{2n} p\left(\frac{-1}{z_2}\right) \log z_2$$

with

$$p(z_1) = \sum_{\mu=0}^{2n} a_\mu z^\mu \,, \quad a_\mu \in \mathbb{C},$$

represents a globally single-valued solution of (208) in K. Moreover, the result is generalized to (m_1,m_2)-circular regions (cf.[27], Theorem 1 and Theorem 2).

Apart from the holomorphic solutions in [5], Ch.3, the properties of the meromorphic solutions of (208) were investigated. Here, we point out a result which characterizes the polar set.

Theorem 34

Let $D^{(k)}$, $k = 1,2$, be a finite (not necessarily simply connected) domain of the z_k-plane and $D = D^{(1)} \times D^{(2)}$.

a) For every solution of (208) meromorphic in D there exist two in $D^{(k)}$ meromorphic generators $g_k(z_k)$, such that

$$w = E_1g_1 + E_2g_2. \tag{210}$$

If $z_k^{(\mu_k)}$, $\mu_k = 1, \ldots, m_k$, $k = 1,2$, are the poles of the generators $g_k(z_k)$ of order s_{μ_k} in $D^{(k)}$, then, the planes $z_k = z_k^{(\mu_k)}$ parallel to the complex coordinate directions represent carriers of poles [11] of order $s_{\mu_k} + n$. Moreover, at most the surface S may appear as a carrier of poles of order n.

b) Conversely, for arbitrary functions $g_k(z_k)$ meromorphic in $D^{(k)}$ (210) represents a meromorphic solution of (208) in D.

c) If $D \cap S = S^* \neq \emptyset$, the solution (210) is holomorphic in all points

$$(z_1^*, z_2^*) \in S^* \text{ with } z_k \neq z_k^{(\mu_k)}, \quad k = 1,2,$$

if, and only if, the generators satisfy the condition

$$g_1(z_1) = (-1)^n z_1^{2n} g_2\left(\frac{-1}{z_1}\right).$$

We wish to point out that the above-mentioned invariance properties of the differential equation (208) permit the construction of automorphic solutions analogous to the procedure applied in the case of the differential equation (92) by E. Peschl (cf.[5], Ch.4).

Proceeding from (208) by the transition to m independent complex variables we obtain the differential equation

(211) $$\omega^2 \Delta_{2m} w + \varepsilon n(n+1) w = 0, \quad n \in \mathbb{N}, \quad \varepsilon = \pm 1,$$ [12]

---

[11] Here, we suppose that possibly isolated ambiguous points may appear.

[12] In [67] G. Jank investigated the representation of solutions of the differential equation

$$\Delta_{2m} w + \varepsilon n(n+1) B(r) w = 0$$

with

$$B(r) = \frac{4\alpha^2 r^{2\alpha-2}}{(1+\varepsilon r^{2\alpha})^2}, \quad r = \left(\sum_{k=1}^{m} z_k \overline{z_k}\right)^{\frac{1}{2}}, \quad \varepsilon = \pm 1, \quad n \in \mathbb{N}, \quad \alpha > 0,$$

by integral operators. The representation can be converted into a form free of integrals if the operator, acting on solutions of the differential equation $\Delta_{2m} w = 0$, has a polynomial kernel.

with

$$z_k = x_k + iy_k, \quad k = 1, \ldots, m,$$

$$\omega = 1 + \varepsilon \sum_{k=1}^{m} z_k \overline{z_k}, \quad \triangle_{2m} = \sum_{k=1}^{m} \frac{\partial^2}{\partial z_k \partial \overline{z_k}},$$

which is closely related to the differential equation of the surface harmonics (cf. Chapter II,2) and was treated in [9]. In this paper it is shown that we may get solutions of (211) defined in polydomains of the space $\mathbb{C}^m$ by use of differential operators, where the generators are solutions of the differential equation

(212) $$\triangle_{2m} h = 0.$$

We set

$$z = (z_1, z_2, \ldots, z_m), \quad \bar{z} = (\overline{z_1}, \ldots, \overline{z_m}),$$

and denote by $D^{(k)}$ a finite simply connected domain of the z_k-plane, $k = 1, 2, \ldots, m$. Moreover, we use

$$D = D^{(1)} \times \ldots \times D^{(m)} \subset \mathbb{C}^m.$$

We denote by D_m the differential operator

$$D_m = \sum_{k=1}^{m} \left\{ z_k \frac{\partial}{\partial z_k} + \overline{z_k} \frac{\partial}{\partial \overline{z_k}} \right\}$$

and suppose that ω does not vanish in D. Then, it follows that every solution of (211) defined in the polydomain D satisfies the differential equation

(213) $$\omega \triangle_{2m}^{r+1}(\omega^r w) = \varepsilon[r(r+1)-n(n+1)] \triangle_{2m}^{r}(\omega^{r-1} w)$$

for $r = 0, 1, \ldots, n$. This assertion is proved by induction, considering the relation

(214) $$\triangle_{2m} D_m = D_m \triangle_{2m} + 2\triangle_{2m}.$$

By r = n in (213) we get

(215) $$\triangle_{2m}^{n+1}(\omega^n w) = 0.$$

The set of the solutions of this differential equation contains the solutions of (211). If we denote by $H^s(D)$ the set of the solutions of the differential equation $\triangle_{2m}^{s} w = 0$, $s \in \mathbb{N}$, defined in D, then

$$\triangle_{2m}^{n}(\omega^n w) = \varphi(z,\bar{z}), \quad \varphi \in H^1(D).$$

Setting

$$W = \psi_1 + \psi_2\omega, \quad \psi_1,\psi_2 \in H^1(D),$$

we get

$$\triangle_{2m} W = \varepsilon D_m \psi_2 + \varepsilon m \psi_2$$

and by (214)

$$\triangle_{2m}^{2} W = 0,$$

to say $W \in H^2(D)$. This suggested to apply

$$W = \sum_{k=0}^{n} h_k(z,\bar{z})\omega^k, \quad h_k(z,\bar{z}) \in H^1(D)$$

to a solution of the differential equation

$$\triangle_{2m}^{n+1} W = 0, \quad n \in \mathbb{N}_0,$$

and

(216) $$w = \sum_{k=0}^{n} \frac{h_k(z,\bar{z})}{\omega^{n-k}}$$

to a solution of (211). Substituting (216) into (211), we obtain the following result (cf.[9], p.11).

Theorem 35

Let $h(z,\bar{z})$ be a solution of the differential equation $\Delta_{2m}h = 0$ defined in the polydomain $D = D^{(1)} \times \ldots \times D^{(m)}$ of the space $\mathbb{C}^m$. Then,

$$w = \sum_{k=0}^{n} \frac{(2n-k)!}{k!(n-k)!\omega^{n-k}} (D_m+m-n-1)_k h \tag{217}$$

represents a solution of (211) in D.

Now we set

$$\tilde{\Delta}_{2m} = \sum_{k=1}^{m} R_k S_k, \quad R_k = \frac{1}{\varphi_k'} \frac{\partial}{\partial z_k}, \quad S_k = \frac{1}{\overline{\psi_k'}} \frac{\partial}{\partial \overline{z}_k},$$

where the functions $\varphi_k(z_k)$, $\psi_k(z_k)$ are holomorphic in $D^{(k)}$ and satisfy the condition

$$\varphi_k'\psi_k' \neq 0, \quad k = 1, \ldots, m.$$

Generalizing the procedure considered in Chapter I,3, we may also map solutions of the differential equation

$$\tilde{\Delta}_{2m}h = 0$$

defined in polydomains of the space $\mathbb{C}^m$ onto solutions of other differential equations by differential operators. In this connection we point out two results proved in [14].

Theorem 36

Let $h(z,\bar{z})$ be a solution of the differential equation $\tilde{\Delta}_{2m}h = 0$ defined in the polydomain $D = D^{(1)} \times \ldots \times D^{(m)}$ of the space $\mathbb{C}^m$. Then,

$$w = \omega^{n+1}\tilde{\Delta}_{2m}^{n} \frac{h}{\omega}, \quad n \in \mathbb{N}, \tag{218}$$

with

$$\omega = \sum_{k=1}^{m} (\varphi_k + \bar{\psi}_k) \neq 0 \text{ in } D$$

represents a solution of the differential equation

$$\omega^2 \tilde{\Delta}_{2m} w - nm(n+1)w = 0$$

in D.

Theorem 37

Let $h(z,\bar{z})$ be a solution of the differential equation $\tilde{\Delta}_{2m} h = 0$ in the polydomain D of the space $\mathbb{C}^m$.
Then,

$$w = \sigma^{\mu+1} \tilde{\Delta}_{2m}^{\mu} \frac{h}{\sigma}, \qquad \mu \in \mathbb{N}, \tag{219}$$

with

$$\sigma = 1 + \sum_{k=1}^{m} \varphi_k \bar{\psi}_k \neq 0 \text{ in } D$$

represents a solution of the differential equation

$$\sigma^2 \tilde{\Delta}_{2m} w + \mu(\mu+1)w = 0$$

in D.

If we use, for instance, $\varphi_k = \psi_k$ in Theorem 37, real-valued solutions may appear. Moreover, if we set

$$\mu = n+m-1, \quad n \in \mathbb{N},$$

and

$$\varphi_k = \psi_k = z_k, \qquad k = 1, \ldots, m,$$

by

$$\sigma^2 \tilde{\Delta}_{2m} w + (n+m-1)(n+m)w = 0 \tag{220}$$

we obtain the differential equation of the surface harmonics of degree n (cf. Chapter I,2).

9) Differentialoperators on solutions of the heat equation

Since there are many parallels between elliptic and parabolic equations, we may expect that we can map also solutions of simple parabolic differential equations onto solutions of other differential equations of this type by means of differential operators. Corresponding investigations can be found in [17], where, analogous to the method in Chapter I,3, solutions of the heat equation are used as generators.[13)]

In this section D denotes a domain of the space $\mathbb{R}^{m+1}$ and $u(x_1,\dots,x_m,t)$ is an arbitrary solution [14)] of the heat equation

$$\Delta u = u_t, \quad \Delta = \sum_{s=1}^{m} \frac{\partial^2}{\partial x_s^2} \tag{221}$$

in D. $\sigma = \alpha(t)\,\beta(\eta)$ with

$$\eta = \sum_{s=1}^{\tau} x_s, \quad 1 \leq \tau \leq m,$$

denotes a particular nonvanishing solution of (221) in D. Then,

$$v_o = \frac{u}{\sigma}$$

satisfies the differential equation

$$\Delta v_o + \frac{2}{\sigma} \sum_{s=1}^{\tau} \sigma_{x_s} v_{o,x_s} = v_{o,t} \,. \tag{222}$$

---

13) The mapping of solutions of the heat equation onto solutions of other parabolic equations by integral operators were treated in [39] and [108] by D.Colton and W.Watzlawek respectively. Moreover, in [108] the classification of the operators of the type P respectively P_o, introduced by E.Kreyszig (cf.[79-81]), could be carried over to parabolic and hyperbolic equations.

14) Here and in the following a solution u of (221) in D is a function defined in D which has continuous derivatives $u_t, u_{x_s}, u_{x_s x_s}$, $s=1,\dots,m$, and satisfies the differential equation (221) in D. Such a solution is an infinitely differentiable function (cf. e.g.[53]).

If we apply the operator

$$d_\tau = \sum_{s=1}^{\tau} \frac{\partial}{\partial x_s}$$

to (222), by

$$d_\tau \Delta = \Delta d_\tau$$

it follows the differential equation

$$\Delta v_1 + \frac{2}{\sigma} \sum_{s=1}^{\tau} \sigma_{x_s} v_{1,x_s} + 2Bv_1 = v_{1,t}$$

with

$$B = \left(\frac{d_\tau \sigma}{\sigma}\right)_{x_s} = \tau \, \frac{\beta\beta'' - \beta'^2}{\beta^2} \tag{223}$$

and $v_1 = d_\tau v_o$. In general, if we use

$$v_k = \left[d_\tau - \frac{(k-1) d_\tau \sigma}{\sigma} \right] v_{k-1} , \quad k = 1,2, \ldots, n,$$

we can show by induction that

$$v_n = \left[d_\tau - \frac{(n-1) d_\tau \sigma}{\sigma} \right] \ldots \left[d_\tau - \frac{d_\tau \sigma}{\sigma} \right] d_\tau v_o \tag{224}$$

satisfies the differential equation

$$\Delta v_n + \frac{2}{\sigma} \sum_{s=1}^{\tau} \sigma_{x_s} v_{n,x_s} + n(n+1) B v_n = v_{n,t} \, .$$

Finally, it follows that $v = \sigma v_n$ is a solution of the differential equation

$$\Delta v + n(n+1) B v = v_t \, .$$

In the representation (224) the coefficient $\alpha(t)$ falls away and we get

$$v = \beta\left[d_\tau - \frac{\tau(n-1)\beta'}{\beta}\right] \cdots \left[d_\tau - \tau\frac{\beta'}{\beta}\right] d_\tau \frac{u}{\beta}.$$

Since the function $\beta = \beta(\eta)$ satisfies the ordinary differential equation

(225) $$\tau\beta'' - \lambda\beta = 0, \qquad \lambda \in \mathbb{R},$$

we obtain in detail ($a_k \in \mathbb{R}$):

(226) $$\beta_1 = a_1 + a_2\beta \text{ for } \lambda = 0,$$

(227) $$\beta_2 = a_3\cosh\left(\eta\sqrt{\frac{\lambda}{\tau}}\right) + a_4\sinh\left(\eta\sqrt{\frac{\lambda}{\tau}}\right) \quad \text{for } \lambda > 0,$$

(228) $$\beta_3 = a_5\sin\left(\eta\sqrt{\frac{-\lambda}{\tau}}\right) + a_6\cos\left(\eta\sqrt{\frac{-\lambda}{\tau}}\right) \quad \text{for } \lambda < 0$$

and the corresponding coefficients B:

(229) $$B_1 = \frac{-\tau a_2^2}{(a_1+a_2\eta)^2},$$

(230) $$B_2 = \frac{\lambda(a_3^2-a_4^2)}{\left[a_3\cosh\left(\eta\sqrt{\frac{\lambda}{\tau}}\right)+a_4\sinh\left(\eta\sqrt{\frac{\lambda}{\tau}}\right)\right]^2},$$

(231) $$B_3 = \frac{\lambda(a_5^2+a_6^2)}{\left[a_5\sin\left(\eta\sqrt{\frac{-\lambda}{\tau}}\right)+a_6\cos\left(\eta\sqrt{\frac{-\lambda}{\tau}}\right)\right]^2}.$$ [15]

Theorem 38

Let $u = u(x_1, \ldots, x_m, t)$ be an arbitrary solution of the differential equation (221)

[15] Cf. [108], example 5,2.

$$\triangle u = u_t$$

in D. Let $\beta = \beta(\eta)$,

$$\eta = \sum_{s=1}^{\tau} x_s , \quad 1 \leq \tau \leq m,$$

be a nonvanishing solution of the differential equation

$$\tau \beta'' - \lambda\beta = 0, \quad \lambda \in \mathbb{R},$$

and

$$d_\tau = \sum_{s=1}^{\tau} \frac{\partial}{\partial x_s} .$$

Then,

(232) $$v = \beta\left[d_\tau - \tau \frac{(n-1)\beta'}{\beta}\right] \ldots \left[d_\tau - \tau \frac{\beta'}{\beta}\right] d_\tau \frac{u}{\beta}$$

represents a solution of the differential equation

(233) $$\triangle v + n(n+1) \frac{\lambda\beta^2 - \tau\beta'^2}{\beta^2} v = v_t , \quad n \in \mathbb{N},$$

in D.

Moreover, by induction on n it follows that the representation (232) can be written also in the form

(234) $$v = \sum_{k=0}^{n} p_k(\eta) d_\tau^k u$$

with

$$p_k(\eta) = \sum_{s=0}^{\left[\frac{n-k}{2}\right]} (-1)^{n-k-s} a_{ks} (\tau\lambda)^s \left(\frac{\tau\beta'}{\beta}\right)^{n-k-2s}, \quad a_{ks} > 0.$$ [16]

[16] [m] denotes the largest integer $\leq$ m.

For $\lambda \neq 0$ the coefficients p_k are related to certain inhomogeneous Legendre equations. If we use

$$p_k(\eta) = q_k(y), \qquad y = \frac{\beta'}{\beta}\sqrt{\frac{\tau}{\lambda}},$$

it follows, substituting (234) into (233),

$$(y^2-1)q_k'' + 2yq_k' - n(n+1)q_k = \frac{2}{\sqrt{\lambda\tau}}\, q_{k-1}'$$

for $k = 0,1, \ldots, n$ with $q_{-1} \equiv 0$. In the case $\lambda = 0$ the polynomials reduce by suitable normalization to

$$p_k(\eta) = \frac{(-1)^{n-k}(2n-k)!}{2^{n-k}k!(n-k)!}\left(\frac{\tau\beta'}{\beta}\right)^{n-k}.$$

Moreover, if we imploy the function β in the normalized form $\beta = \eta$, we get the following

Theorem 39

Let $u = u(x_1, \ldots, x_m, t)$ be an arbitrary solution of the differential equation (221)

$$\Delta_u = u_t$$

in D.
Then,

$$v = \sum_{k=0}^{n} \frac{(-\tau)^{n-k}(2n-k)!}{2^{n-k}k!(n-k)!}\,\frac{d_\tau^k u}{\eta^{n-k}}$$

represents a solution of the differential equation

$$\Delta v - \frac{n(n+1)}{\eta^2}\, v = v_t, \qquad n \in \mathbb{N},$$

in D.

Whereas in the above-mentioned differential equations the coefficients of v represents a function of the variables $x_1, \ldots, x_m$, in the following we shall consider differential equations, where this coefficient depends on t. As before u denotes an arbitrary solution of the heat equation (221) defined in D. σ is a nonvanishing particular solution of (221) in D which satisfies the condition

$$\frac{d_\tau \sigma}{\sigma} = a(t) + \eta\, b(t)$$

with suitable functions a(t) and b(t). Then, for

$$w_o = \frac{u}{\sigma}$$

we have

$$\Delta w_o + \frac{2}{\sigma} \sum_{s=1}^{m} \sigma_{x_s} w_{o,x_s} = w_{o,t} ,$$

and by induction it follows that

$$w_n = d_\tau^n w_o , \qquad n \in \mathbb{N},$$

is a solution of the differential equation

$$\Delta w_n + \frac{2}{\sigma} \sum_{s=0}^{m} \sigma_{x_s} w_{n,x_s} + 2nb(t) w_n = w_{n,t}$$

in D. Transforming by

$$w = \sigma w_n ,$$

we obtain

$$\Delta w + 2nb(t) w = w_t .$$

Theorem 40

Let $u = u(x_1, \ldots, x_m, t)$ be an arbitrary solution of (221)

$$\Delta u = u_t$$

in D. Let σ be a particular solution of (221) in D which satisfies the conditions

(235) (i) $\sigma \neq 0$ in D

(236) (ii) $\frac{d_\tau \sigma}{\sigma} = a(t) + \eta b(t).$

Then,

(237) $$w = \sigma d_\tau^n \frac{u}{\sigma}$$

represents a solution of the differential equation

$$\Delta w + 2nb(t)w = w_t, \qquad n \in \mathbb{N},$$

in D.

Employing here, for instance, the well known fundamental solution

$$\sigma = t^{-\frac{m}{2}} e^{-\frac{1}{4t} \sum_{j=1}^{m} x_j^2}$$

of the heat equation, it follows

$$\frac{d_\tau \sigma}{\sigma} = - \frac{\eta}{2t}$$

and

$$\Delta w - \frac{n}{t} w = w_t , \quad n \in \mathbb{N}.$$

First, for the representation (237) we get

$$w = \sigma d_\tau^n \frac{u}{\sigma} = \sum_{k=0}^{n} f_k(\eta, t) d_\tau^k u$$

with

(238) $$f_k(\eta, t) = \binom{n}{k} \left[\frac{\eta}{2t} + \tau \frac{\partial}{\partial \eta} \right]^{n-k} 1$$

and then by induction

$$(239) \qquad f_k(\eta,t) = \frac{n!}{2^{n-k}k!} \sum_{s=0}^{[\frac{n-k}{2}]} \frac{\tau^s}{s!(n-k-2s)!} \frac{\eta^{n-k-2s}}{t^{n-k-s}} .$$

Theorem 41

D denotes a domain of the space $\mathbb{R}^{m+1}$ with $t \neq 0$ in D. Let u be an arbitrary solution of (221)

$$\triangle u = u_t$$

in D.
Then,

$$(240) \qquad w = \sum_{k=0}^{n} f_k(\eta,t) d_\tau^k u$$

with $f_k(\eta,t)$ according to (238) respectively (239) represents a solution of the differential equation

$$(241) \qquad \triangle w - \frac{n}{t} w = w_t$$

in D.

Since we can obtain a differential equation of the form

$$\triangle w + C(t)w = w_t$$

by a suitable transformation from the differential equation (221), here it is possible, proceeding from a known solution of the heat equation, to get further (non-trivial) solutions of this differential equation. In particular, by this procedure we may obtain solutions of the heat equation, in which also the variable t arises, from arbitrary solutions of the Laplace equation $\triangle u = 0$.

If we denote by v a solution of the differential equation

$$\triangle v = v_t ,$$

the function

$$w = \frac{v}{\alpha(t)}, \quad \alpha(t) \neq 0,$$

satisfies the differential equation

$$\Delta w + C(t)w = w_t \quad \text{with} \quad C(t) = -\frac{\alpha'}{\alpha}.$$

Thus, we get by (239) with

$$w = \sum_{k=0}^{n} g_k(\eta, t) d_\tau^k u$$

and

$$g_k(\eta, t) = \frac{n!}{2^{n-k} k!} \sum_{s=0}^{[\frac{n-k}{2}]} \frac{\tau^s t^{k+s}}{s!(n-k-2s)!} \eta^{n-k-2s},$$

again a solution of the heat equation in D if u denotes a solution of (221) in D.

10) Bergman operators with polynomials as generating functions

Another approach to the representation of solutions of partial differential equations by differential operators offers by Bergman integral operators (cf. e.g.[32] and [55]) if we are in a position to convert the representation to a form free of integrals. This is possible in case of Bergman operators with polynomials as generating functions, as was shown in [76] by M.Kracht and E.Kreyszig (cf. also [24,48,65,107]. Thus, the question is of interest for which classes of differential equations such Bergman operators exist. The first criteria of this kind for Bergman operators to a single holomorphic function were derived by E.Kreyszig [79] and M.Kracht [74]. A further systematic treatment which includes also the consideration of pairs of Bergman operators acting on pairs of holomorphic functions can be found in a paper of M.Kracht [75] (cf. also [77]). From this paper we sketch some investigations and summarize certain results as far as they are of interest in connection with the differential operators considered here.

Let H(D) be the set of functions of the variables z_1 and z_2 which are holomorphic in an open subset $D \subset \mathbb{C} \times \mathbb{C}$. Let M be an open subset of D. Then,

$$(242) \qquad L = \frac{\partial^2}{\partial z_1 \partial z_2} + a_1 \frac{\partial}{\partial z_1} + a_2 \frac{\partial}{\partial z_2} + a_3$$

with

$$a_1,\ a_2,\ a_3 \in H(D)$$

is a linear operator which maps H(D) onto itself.

In the following we consider the differential equation

$$(243) \qquad Lw = 0.$$

It is well known that (243) can be transformed into differential equations in which one of the first derivatives does not appear. If we use the operators

$$(244) \qquad L_j = \frac{\partial^2}{\partial z_1 \partial z_2} + b_j \frac{\partial}{\partial z_{3-j}} + c_j\ , \quad j = 1,2,$$

with

(245)
$$\begin{cases} b_j = a_{3-j} - \int_{\xi_{3-j}}^{z_{3-j}} [a_j(\zeta_1,\zeta_2)\Big|_{\zeta_j=z_j}]_{z_j} d\zeta_{3-j} , \\ \\ c_j = a_3 - a_1 a_2 - a_{j,z_j} , \end{cases}$$

(243) is transformed into

(246)
$$L_j u_j = 0$$

by

$$w = u_j \exp\left[-\int_{\xi_{3-j}}^{z_{3-j}} a_j(\zeta_1,\zeta_2)\Big|_{\zeta_j=z_j} d\zeta_{3-j}\right] .$$

We set $(\xi_1,\xi_2) \in D$ and

$$K_{\rho_j}(\xi_j) = \{ z \in \mathbb{C} \,|\, |z_j-\xi_j| < \rho_j \} , \quad j = 1,2.$$

We denote by $D_{\rho_1,\rho_2}(\xi_1,\xi_2)$ a bicylinder

$$D_{\rho_1,\rho_2}(\xi_1,\xi_2) = K_{\rho_1}(\xi_1) \times K_{\rho_2}(\xi_2) \subset D$$

and by S_j a rectifiable arc in $K_1(0) \cup \partial K_1(0)$ from -1 to $+1$. Let $\tilde{g}_j$ be a holomorphic function of z_1, z_2, and t in $D_{\rho_1,\rho_2}(\xi_1,\xi_2) \times K_1(0)$ and

$$f_j \in H(K_{r_j}(\xi_j)), \quad r_j > 0, \quad j = 1,2,$$

then, by

(247)
$$(B_j f_j)(z_1,z_2) = \int_{S_j \, -1}^{1} \tilde{g}_j(z_1,z_2,t) f_j\left(\xi_j + \frac{z_j-\xi_j}{2}(1-t^2)\right) \frac{dt}{(1-t^2)^{1/2}}$$

we define operators with

$$B_j: \; H(K_{r_j}(\xi_j)) \to H(K_{s_j}(\xi_j) \times K_{\rho_{3-j}}(\xi_{3-j}))$$

and

$$s_j = \min(2r_j, \rho_j), \quad j = 1,2.$$

The function f_j in (247) is termed associated function, $\tilde{g}_j$ is termed Bergman generator and the operator B_j is called Bergman operator if

$$(248)\quad \tilde{g}_j = g_j e_j, \quad e_j(z_1,z_2) = \exp\left[-\int_{\xi_{3-j}}^{z_{3-j}} a_j(\zeta_1,\zeta_2)\Big|_{\zeta_j=z_j} d\zeta_{3-j}\right]$$

with

$$(249)\quad (1-t^2)g_{j,z_{3-j}t} - \frac{1}{t}\, g_{j,z_{3-j}} + 2(z_j-\xi_j)tL_j g_j = 0$$

$$\text{for} \quad (z_1,z_2,t) \in D_{\rho_1,\rho_2}(\xi_1,\xi_2) \times K_1(0),$$

$$(250)\quad [(z_j-\xi_j)t]^{-1} g_{j,z_{3-j}} \quad \text{continuous for } (z_1,z_2,t) \in D_{\rho_1,\rho_2}(\xi_1,\xi_2) \times S_j,$$

$$(251)\quad (1-t^2)^{1/2} g_{j,z_{3-j}} \xrightarrow{\text{uniformly}} 0 \text{ for } t \to \pm 1 \text{ and}$$

$$(z_1,z_2) \in D_{\rho_1,\rho_2}(\xi_1,\xi_2).$$

Using the above notation we get the following theorem (cf.[32]).

Theorem 42

Let $\tilde{g}_j$, $j = 1,2$, be Bergman generators and $f_j \in H(K_{r_j}(\xi_j))$. Then

$$(252)\quad w = \alpha_1 w_1 + \alpha_2 w_2$$

with

$$w_j = (B_j f_j)(z_1,z_2), \quad \alpha_j \in \mathbb{C}, \quad j = 1,2,$$

represents a twice continuously differentiable solution of $Lw = 0$ in

$D_{s_1,s_2}(\xi_1,\xi_2)$.

If in (252) the two coefficients α_1 and α_2 are unequal zero, we have a pair of Bergman operators acting on pairs of holomorphic functions.

A Bergman generator $\tilde{g}_j$ is called polynomial generator of degree n (cf. [79]) if

$$g_j(z_1,z_2,t) = \sum_{\mu=0}^{n} q_{j,2\mu}(z_1,z_2)t^{2\mu} ,$$

(253)

$$g_{j,2n} \neq 0, \quad n \in \mathbb{N}_0 .$$

A polynomial generator is called minimal (cf.[78]) if there exists no polynomial generator of lower degree for the considered differential equation. The corresponding classes of differential operators L are marked as follows (cf.[74] and [81]):

$P_{j,n}$ denotes the class of operators (242) for which there exist polynomial generators $\tilde{g}_j$ of degree n, $n \in \mathbb{N}_0$, for (243).

$P^o_{j,n}$ denotes the class of all $L \in P_{j,n}$ with $L \notin P_{j,m}$ for m=0,1,...,n-1.

Let Λ_i^* be the class of all operators (242) with $a_i = 0$.

The class $P_{j,n}$ may implicitely be characterized as follows. If we substitute g_j according to (253) into (249), we obtain for $(\xi_1,\xi_2)=(0,0)$ and b_j, c_j according to (245) the following system of linear partial differential equations for the functions $g_{j,2\mu}$

$$(2\mu-1)q_{j,2\mu;z_{3-j}} - 2(\mu-1)q_{j,2\mu-2;z_{3-j}} + 2z_j L_j q_{j,2\mu-2} = 0,$$

(254)

$$\mu \in \mathbb{Z}, \quad q_{j,2\mu} = 0 \ \text{ for } \ \mu = -\nu \ \text{ and } \ \mu = n+\nu , \quad \nu \in \mathbb{N},$$

if we consider that the coefficients of each t-power have to vanish. Thus, we get the following assertion (cf. [79]).

Theorem 43

a) $L \in P_{n,j}$ if, and only of, there are functions $s_{j,2\mu}(z_1,z_2)$, $\mu = 1, \ldots, n+1$, such that

$$s_{j,2n+2} = 0,$$

where $s_{j,2n+2}$ are given by the recursive system

$$s_{j,2} = q_{j,o}c_j \, ,$$

$$s_{j,2\mu+2} = M_{j,2\mu}s_{j,2\mu} \, , \quad \mu = 1, \ldots, n,$$

$$M_{j,2}s = \frac{2}{2\mu-1} z_j \left\{ s_{z_j} + [b_j - \frac{\mu-1}{z_j}]s + c_j \int s \, dz_{3-j} \right\} .$$

b) In the case of $L \in P_{j,n}$ it follows for the functions $q_{j,2\mu}$ in (253):

$$q_{j,o}(z_1,z_2) = q_{j,o}(z_j),$$

$$q_{j,2\mu}(z_1,z_2) = \frac{(-1)^{\mu}}{2\mu-1} 2z_j \int s_{j,2\mu}(z_1,z_2)dz_{3-j} \quad \text{for} \quad \mu = 1, \ldots, n.$$

In [76] M.Kracht and E.Kreyszig investigated the question whether it is possible to convert the solutions, given by Bergman operators with polynomial generators according to Theorem 42, to a form free of integrals. From this paper we quote the following fundamental assertion.

Theorem 44

Let $L \in P_{j,n} \cap \Lambda_j^*$. Let w_j be a solution of (243) according to (247). Let $\tilde{g}_j$ be a polynomial generator of degree n and $f_j \in H(K_{r/2}(0))$. Here, r denotes the radius of the largest bicylinder $D_{r,r}(0,0) \subset D$. The integration path goes from $t = -1$ to $t = +1$ along the real axis. Then,

$$w_j = \tilde{w}_j$$

with

$$\tilde{w}_j(z_1,z_2) = (\tilde{B}_j\tilde{f}_j)(z_1,z_2),$$

(255) $$\tilde{B}_j = \sum_{\mu=0}^{n} \frac{(2\mu)!}{2^{2\mu}\mu!} \tilde{q}_{j,2\mu}(z_1,z_2) \frac{d^{n-\mu}}{dz_j^{n-\mu}},$$

if the coefficients $q_{j,2\mu}$ of the polynomial generators $\tilde{g}_j$ and the coefficients $\tilde{q}_{j,2\mu}$ in (255) satisfy the relation

$$\tilde{q}_{j,2\mu}(z_1,z_2) = z_j^{-\mu} q_{j,2\mu}(z_1,z_2),$$

and if the relations

$$\delta_{j,\nu} = \begin{cases} 0 \text{ for } \nu < n, \\ \frac{[2(\nu-n)]!\pi}{2^{3(\nu-n)}(\nu-n)!\nu!} \gamma_{j,\nu-n} \text{ for } \nu \geq n \end{cases}$$

are valid for the coefficients $\gamma_{j,\nu}$ of the expansion of the associated functions f_j of w_j about the origin and the coefficients $\delta_{j,\nu}$ of the expansion of the associated functions $\tilde{f}_j$ of $\tilde{w}_j$.

The assertion in Theorem 43 about the existence of Bergman operators with polynomial generators is very complicated in regard to the practical application. Further criteria and possibilities for the construction of differential equations and the corresponding polynomial generators were derived by M.Kracht in [75] by means of the theory of the Laplace invariants (cf.[40]). In the following these results are briefly summarized.
We set

(256) $$L^{(0)}w = w_{z_1z_2} + a_{10}w_{z_1} + a_{20}w_{z_2} + a_{30}w = 0$$

and

(257) $$h_{j,\rho} = a_{j\rho,z_j} + a_{1\rho}a_{2\rho} - a_{3\rho}, \quad j = 1,2, \quad \rho \in \mathbb{N}_0.$$

Here, the index o is added at the operator L and the coefficients a_1, a_2, a_3, since subsequently certain transformations will be applied to (256). Let T_j, $T_{j,\rho}$ be the operators

(258) $$T_j = \frac{\partial}{\partial z_{3-j}} + a_{j_o}, \quad T_{j,\rho} = \frac{\partial}{\partial z_{3-j}} + a_{j,\rho-1}, \quad \rho \geq 2.$$

Then, in case $h_{j,o} = 0$ by

(259) $$W_{j,\rho} = T_j^{(\rho)} w, \quad T_j^{(\rho)} = T_{j,\rho} \cdots T_{j,2} T_j$$

it follows

(260) $$L^{(0)} w = W_{j,1;z_j} + a_{3-j,0} W_{j,1} = 0.$$

The differential equation (260) can be integrated; first, we get

$$W_{j,1}(z_1,z_2) = \exp\left\{-\int a_{3-j,0} dz_j\right\} F_{3-j}(z_{3-j})$$

and then

(261) $$w(z_1,z_2) = \left\{F_j(z_j) + \int F_{3-j}(z_{3-j}) \exp\left[\int (a_{j_o} dz_{3-j} - a_{3-j,0} dz_j)\right] dz_{3-j}\right\} \exp\left[-\int a_{j_o} dz_{3-j}\right].$$

If $F_1(z_1)$ amd $F_2(z_2)$ denote arbitrary holomorphic functions, (261) is a general solution of $L^{(0)} w = 0$ with $h_{j,o} = 0$, $j = 1$ or $j = 2$. Since this result can be derived in case $h_{j,o} = 0$, in [75] such differential equations were considered for which we get a differential equation with $h_{j,o} = 0$, only if we apply n times the operator (258).
Let $L^{(\rho)}$ be the differential operator

(262) $$L^{(\rho)} = \frac{\partial^2}{\partial z_1 \partial z_2} + a_{1\rho} \frac{\partial}{\partial z_1} + a_{2\rho} \frac{\partial}{\partial z_2} + a_{3\rho}.$$

We use $h_{j,\rho}$ and $W_{j,\rho}$ according to (257) and (259) respectively. If $h_{j,\rho} \neq 0$ for $\rho = 1, \ldots, n-1$, by $L^{(0)} w = 0$ it follows

$$L^{(n)} W_{j,n} = 0$$

with

$$(263)\quad \begin{cases} a_{jn} = a_{j0} - \left[\log \prod_{\rho=0}^{n-1} h_{j,\rho}\right]_{z_{3-j}} \\ a_{3-j,n} = a_{3-j,0} \\ a_{3n} = a_{jn,z_j} + a_{1n}a_{2n} - h_{j,n} \end{cases}$$

and

$$(264)\quad \begin{cases} h_{j,n} = (n+1)h_{j,0} - nh_{3-j,0} - \left[\log \prod_{\rho=0}^{n-1} h_{j,\rho}^{n-\rho}\right]_{z_1 z_2} \\ h_{3-j,n} = h_{j,n-1} = nh_{j,0} - (n-1)h_{3-j,0} - \left[\log \prod_{\rho=0}^{n-2} h_{j,\rho}^{n-1-\rho}\right]_{z_1 z_2} . \end{cases}$$

Using these relations, it is possible to obtain another criterion for the appearance of polynomial generators which we quote in the following (cf. [75], p.64).

Theorem 45

For the differential equation (256), $L^{(0)}w = 0$, there exists a minimal polynomial generator $\tilde{g}_j$ of degree n for a Bergman operator to a single holomorphic function (that is $L^{(0)} \in P^0_{j,n}$) if, and only if, by the transformation $T_j^{(n)}$ to w we get an equation $L^{(n)}W_{j,n} = 0$ with $h_{j,n} = 0$ and $h_{j,\rho} \neq 0$ for $\rho = 1, \ldots, n-1$.

If the conditions in Theorem 45 are satisfied for $j = 1$ (with $n \in \mathbb{N}$) and also for $j = 2$ (with $m \in \mathbb{N}$), it follows a corresponding assertion about the existence of polynomial generators for a pair of Bergman operators acting on pairs of holomorphic functions. Moreover, in [75] by means of Theorem 45 another important assertion was proved for pairs of Bergman operators (cf. [75], p.74).

Theorem 46

For $L^{(0)} \in P^0_{j,n}$, $n \in \mathbb{N}_0$, it follows $L^{(0)} \in P^0_{j,n} \cap P_{3-j,m}$, $m \in \mathbb{N}_0$, if, and only if, $\exp \gamma_{j,n}$ with

$$\gamma_{j,n} = \int (a_{j,n} dz_{3-j} - a_{3-j,n} dz_j)$$

satisfies an ordinary differential equation of the form

$$(265) \qquad \sum_{\rho=0}^{n+m} (-1)^{\rho} \frac{\partial^{\rho}}{\partial z_{3-j}^{\rho}} \delta_{j,\rho} e^{\gamma_{j,n}} + (-1)^{n+m+1} \frac{\partial^{n+m+1}}{\partial z_{3-j}^{n+m+1}} e^{\gamma_{j,n}} = 0$$

where $\delta_{j,\rho}$, $\rho = 0,1, \ldots, n+m$, are functions which depend only on z_{3-j}.

Proceeding from the differential equation (265) the proof of Theorem 46 yields also a procedure for the construction of differential operators of the type $L^{(0)} \in P^{o}_{j,n} \cap P_{3-j,m}$. In [75] the applicability was demonstrated deriving the differential equation (56) and the representation of its solutions considered in Theorem 6. Here, it is sufficient to start from the differential equation (265) in the special form

$$\frac{\partial^{n+m+1}}{\partial z_{3-j}^{n+m+1}} e^{\gamma_{j,n}} = 0, \qquad n,m \in \mathbb{N}_0.$$

Moreover, proceeding from the ordinary differential equation

$$(266) \qquad \sum_{\rho=0}^{n+m} \delta_{j,\rho}(z_{3-j}) y_j^{(\rho)} + y_j^{(n+m+1)} = 0$$

in [75] a second possibility for the construction of differential equations, representations of the solutions, and polynomial generators was derived, where the solution is obtained in form of a (n+m+1)-rowed determinant. The applicability of this procedure was demonstrated in [75], starting from the differential equation (266) in the form

$$y_1^{(2n+1)} = 0$$

and deriving the representation for the solutions of the differential equation

$$(z_1+z_2)^2 w_{z_1 z_2} - n(n+1)w = 0$$

(cf. Theorem 9).

In comparison with Theorem 43 the criteria in Theorem 45 and Theorem 46 are of special importance since no integro-differential-equations have to be solved. However, even in case of simple differential equations the application of these criteria becomes rather complicated if we try to find out whether there exist polynomial generators for a given differential equation and by it representations of the solutions by differential operators. This becomes evident by a corresponding investigation in the case of the differential equation (165) considered in Chapter I,5. The same is true for the application of the principles of constructions resulting from these criteria if representations for the solutions by differential operators are to be derived explicitly. Here, the required labour in calculation could be reduced considerably if it would be possible to derive corresponding criteria in which, in place of holomorphic functions of z_1 or z_2, solutions of simple differential equations of the form (243) are mapped into solutions of more complicated differential equations of this type (cf. Theorem 15 - Theorem 24).

11) Vekua operators

In [102] I.N. Vekua treated the representation of solutions of elliptic differential equations by means of integral operators using the Riemann function. In special cases these representations of solutions may be converted to a form free of integrals by integration by parts. By this we get representations of solutions by differential operators.

The results proved by I.N. Vekua were used by several mathematicians in order to deduce relations with the differential operators considered here (cf. e.g.[58-62, 65, 92, 93]).

R. Heersink used in [59, 61, 62] the results derived in [102] to characterize the differential operators and to investigate the function theoretic properties of the solutions. In [59], for instance, the differential equation (67) was considered, the Riemann function was determined, the solutions were converted in an integral-free form, and the relation between this representation and that one of Theorem 9 was derived.

In addition, in [59] a representation of a particular solution of the inhomogeneous differential equation (84) was given by means of the Riemann function. Using this representation in [60], among other things, the result of Theorem 11 was deduced.

Moreover, applying Vekua operators, R. Heersink proved in [61] an important sufficient criterion for the representation of solutions by differential operators which is quoted in the following.

Theorem (R. Heersink)

1) Let D_1 and D_2 be simply connected domains of the complex plane.
2) Let T_1 and T_2 be differential operators of the form

$$T_1 = \sum_{k=0}^{n_1} a_{1,k} \frac{\partial^k}{\partial z^k} \quad \text{and} \quad T_2 = \sum_{k=0}^{n_2} a_{2,k} \frac{\partial^k}{\partial z_2^k} ,$$

where the functions $a_{j,k}(z_1,z_2)$ are holomorphic in $D_1 \times D_2$ for $k = 0,1, \ldots, n_j$ and $a_{j,n_j} \neq 0$ in $D_1 \times D_2$, $j = 1,2$.

3) Let $g_j(z_j)$ be holomorph in D_j, $j = 1,2$, and let $T_j g_j$ be a solution of the differential equation

(267)
$$w_{z_1 z_2} + A_1 w_{z_1} + A_2 w_{z_2} + A_3 w = 0$$

in $D_1 \times D_2$, where the coefficients $A_s(z_1,z_2)$, $s = 1,2,3$, are holomorphic in $D_1 \times D_2$.

Then, for every solution w of (267), holomorphic in $D_1 \times D_2$, there exist two generators $g_1(z_1)$ and $g_2(z_2)$ holomorphic in D_1 and D_2 respectively, such that

(268)
$$w = T_1 g_1 + T_2 g_2.$$

W. Watzlawek [105] could show that the concept of fundamental systems of solutions of ordinary differential equations may be generalized to partial differential equations if these equations permit a representation of solutions in the form (268). In [106] relations were investigated between generalized fundamental systems, the Riemann function, and Bergman operators.

The Riemann function for the differential equation

(269)
$$w_{z_1 z_2} + \frac{n-m}{\tau} \tau_{z_1} w_{z_2} + \frac{n(m+1)}{\tau^2} w = 0$$ [17)]

with $n,m \in \mathbb{C}$ and

$$\tau = az_1 z_2 + bz_1 + cz_2 + d \neq 0, \quad a,b,c,d \in \mathbb{C}, \quad ad-bc = 1,$$

was determined by J. Püngel [93]. In addition, in [93] differential operators were considered by which solutions of the differential equation

---

17) Proceeding from $(\varphi_1+\varphi_2)^2 w_{z_1 z_2} + (n-m)\varphi_1'(\varphi_1+\varphi_2) w_{z_2} - n(m+1)\varphi_1'\varphi_2' w = 0$ with $\varphi_k(z_k)$ holomorphic in D_k, $\varphi_1+\varphi_2 \neq 0$ (cf. (56)), we get (269), for instance, with $\varphi_1 = z_1$, $\varphi_2 = (cz_2+d)(az_2+b)^{-1}$.

$$w_{z_1 z_2} + A_2 w_{z_2} + A_3 w = 0$$

are mapped onto solutions of equations of the same type. In this context especially operators of the form

$$e^{\alpha_1} R^n e^{\alpha_2}, \quad e^{\alpha_1} S^n e^{\alpha_2},$$

with $\alpha_k(z_1,z_2)$ holomorphic in $D_1 \times D_2$, $k = 1,2$, $R = \frac{\partial}{\partial z_1}$, $S = \frac{\partial}{\partial z_2}$ were treated. Using the above-mentioned theorem of R. Heersink, it was shown that all solutions of the differential equation (269), $n,m \in \mathbb{N}$, defined in $D_1 \times D_2$ are given by

$$w = \tau^{m+1}[R^n(\tau^{-m-1} g_1(z_1)) + S^m(\tau^{-n-1} g_2(z_2))],$$

$$g_k(z_k) \text{ holomorphic in } D_k,\ k = 1,2.$$

Moreover, in [93] the Riemann function of the generalized Darboux equation (cf. (165))

$$w_{z_1 z_2} + w \sum_{k=1}^{4} \frac{n_k(n_k+1)}{\tau_k^2} = 0 \tag{270}$$

with

$$n_k \in \mathbb{C}, \quad \tau_k = a_k z_1 z_2 + b_k z_1 + c_k z_2 + d_k,$$

$$a_j d_k - b_j c_k + a_k d_j - b_k c_j = 2\delta_{jk},$$

was determined. This Riemann function is a generalized hypergeometric function of four variables. In case $n_k \in \mathbb{N}$, $k = 1,2,3,4$, it was possible to convert the solutions of (270) in a form free of integrals and to give a representation by means of differential operators.

CHAPTER II

Applications

1) Spherical surface harmonics and hyperboloid functions

Every solution of the differential equation

(1) $$(1+z\bar{z})^2 w_{z\bar{z}} + n(n+1)w = 0, \quad n \in \mathbb{N},$$

which is defined on the whole Riemann number sphere can be represented in the form (cf. Theorem I,14)

$$w = E_n g$$

with

$$g(z) = \sum_{\mu=0}^{2n} c_\mu z^\mu, \qquad c_\mu \in \mathbb{C},$$

E_n according to (I,98) and $\varepsilon = +1$. For every real-valued solution of (1) defined on the whole Riemann number sphere there exist n+1 complex-valued constants c_μ, $\mu = 0,1, \ldots, n$, such that

(2) $$w = E_n g + \overline{E_n g} \text{ with } g(z) = \sum_{\mu=0}^{n} c_\mu z^\mu$$

(cf. Theorem I,13). The totality of all these real-valued functions is identical with the spherical surface harmonics. If we transform equation (1) by

(3) $$w(z,\bar{z}) = Y(\vartheta,\varphi), \qquad z = \frac{\sin\vartheta}{1-\cos\vartheta}\, e^{i\varphi},$$

$$0 \leq \varphi < 2\pi, \quad 0 \leq \vartheta \leq \pi.$$

we get the known differential equation of the spherical surface harmonics [87]

$$\frac{1}{\sin\vartheta}\frac{\partial}{\partial\vartheta}\left(\sin\vartheta\,\frac{\partial Y}{\partial\vartheta}\right) + \frac{1}{\sin^2\vartheta}\,\frac{\partial^2 Y}{\partial\varphi^2} + n(n+1)Y = 0,$$

and in place of (2) we obtain

$$(4) \qquad Y_n = \sum_{m=0}^{n} \{A_{nm}\cos(m\varphi) + B_{nm}\sin(m\varphi)\} P_n^m(\cos\vartheta),$$

$$A_{nm},\ B_{nm} \text{ constant.}$$

Here we use, as usual,

$$P_n^m(\cos\vartheta) = (-1)^m \sin^m\vartheta \left[\frac{d^m P_n(x)}{dx^m}\right]_{x=\cos\vartheta},$$

where $P_n(x)$ denote the Legendre polynomials. The tesseral harmonics are contained in (4) with

$$Y_{nm}(\vartheta,\varphi) = \begin{cases} \cos(m\varphi)P_n^m(\cos\vartheta) \\ \sin(m\varphi)P_n^m(\sin\vartheta) \end{cases} \qquad m = 0, 1, \ldots, n.$$

We get these functions by the representation (2) with the generators (cf. [4], Ch. IV,2)

$$(5) \qquad g(z) = \frac{(-1)^{n+m}}{2(n-m)!} z^{n-m}$$

and

$$(6) \qquad g(z) = \frac{i(-1)^{n+m}}{2(n-m)!} z^{n-m} .$$

The zonal and sectorial harmonics are contained in (5) and (6) with $m = 0$ and $m = n$ respectively. If we substitute these generators into (2), we obtain by

$$B_s = \frac{n!(-1)^{s+n}}{(n-m)!} \binom{m+n+s}{n} \binom{n-m}{s}$$

the following result.

Tesseral harmonics

Generators: $g(z) = \frac{(-1)^{n+m}}{2(n-m)!} z^{n-m}$ and $g(z) = \frac{i(-1)^{n+m}}{2(n-m)!} z^{n-m}$,

$$(7) \qquad w_{nm} = \begin{cases} \dfrac{z^m + \bar{z}^m}{2(1-z\bar{z})^m} \displaystyle\sum_{s=0}^{n-m} B_s \left(\frac{z\bar{z}}{1+z\bar{z}}\right)^s \\ \\ \dfrac{z^m - \bar{z}^m}{2i(1+z\bar{z})^m} \displaystyle\sum_{s=0}^{n-m} B_s \left(\frac{z\bar{z}}{1+z\bar{z}}\right)^s \end{cases} \qquad m = 0, 1, \ldots, n.$$

Zonal harmonic

Generator: $g(z) = \frac{(-1)^n}{2n!} z^n$,

$$(8) \qquad w_{no} = \sum_{s=0}^{n} B_s \left(\frac{z\bar{z}}{1+z\bar{z}}\right)^s .$$

Sectorial harmonics

Generators: $g(z) = \frac{1}{2}$ and $g(z) = \frac{i}{2}$,

$$(9) \qquad w_{nn} = \begin{cases} \dfrac{z^n + \bar{z}^n}{2(1+z\bar{z})^n} \dfrac{(2n)!(-1)^n}{n!} \\ \\ \dfrac{z^n - \bar{z}^n}{2i(1+z\bar{z})^n} \dfrac{(2n)!(-1)^n}{n!} . \end{cases}$$

Let c_1 and c_2 be arbitrary real numbers; let

$$w_k = E_n g_k + cj, \quad k = 1, 2, \text{ }^{18)}$$

18) We use the notation $A + cj := A + \bar{A}$.

be two real-valued solutions of (1) which have a common domain of definition. Then, by the linear character of the differential operator E_n the solution $c_1w_1 + c_2w_2$ may be represented in the form

$$c_1w_1 + c_2w_2 = E_n(c_1g_1 + c_2g_2) + cj.$$

That is, for example, the addition of two spherical surface harmonics of degree n is realized in this representation by the addition of two polynomials of degree n.

Proceeding from the differential equation (I,92), in the case $\varepsilon = -1$ a class of real-valued solutions of the differential equation

$$(1-z\bar{z})^2 w_{z\bar{z}} - n(n+1)w = 0, \quad n \in \mathbb{N}, \tag{10}$$

corresponds to the spherical surface harmonics. These functions may be termed hyperboloid functions. If we transform equation (10) by

$$w(z,\bar{z}) = H(u,v), \quad z = \frac{\sinh u}{1+\cosh u} e^{iv},$$

we obtain the differential equation

$$\frac{1}{\sinh^2 u} \frac{\partial^2 H}{\partial v^2} + \frac{1}{\sinh u} \frac{\partial}{\partial u} \left(\sinh u \frac{\partial H}{\partial u} \right) - n(n+1)H = 0. \tag{11}$$

On the other hand we get this differential equation if we transform the hyperbolic equation

$$V_{x_1x_1} + V_{x_2x_2} - V_{x_3x_3} = 0 \tag{12}$$

by

$$V(x_1,x_2,x_3) = U(u,v,r)$$

$$\begin{cases} x_1 = r \sinh u \cos v \\ x_2 = r \sinh u \sin v \\ x_3 = r \cosh u \end{cases} \qquad \begin{matrix} 0 \leq v < 2\pi \\ -\infty < u,r < \infty \end{matrix}$$

and separate r by

$$U(u,v,r) = f(r)H(u,v).$$

Thus, the considered functions are solutions of the differential equation (12) which are defined on a hyperboloid of revolution. Here, corresponding to Y_{nm} we get the following particular solutions of (11):

$$H_{nm}(u,v) = \begin{cases} \cos(mv)\, P_n^m(\cosh u) \\ \sin(mv)\, P_n^m(\cosh u) \end{cases} \qquad m = 0,1, \ldots, n,$$

where

$$P_n^m(x) = (x^2-1)^{\frac{m}{2}}\, P_n^{(m)}(x)$$

denotes the associated Legendre function of order m. Then, we obtain the general hyperboloid function by

$$H_n(u,v) = \sum_{m=0}^{n} [C_{mn}\cos(mv)+D_{mn}\sin(mv)]\, P_n^m(\cosh u),$$

$$C_{nm},\ D_{nm} \text{ constant.}$$

In the case of the differential equation (10) the corresponding solution is

$$w = E_n g + \overline{E_n g}, \quad g(z) = \sum_{\mu=0}^{n} c_\mu z^\mu , \quad c_\mu \in \mathbb{C},$$

with E_n according to (I,98) and $\varepsilon = -1$. Considering the maximum principle for the solutions of (10) (cf.[4], Theorem 3), it follows that there is no zonal hyperboloid function. Moreover, there are no tesseral functions; here, we have a nonvanishing hyperboloid function (cf.[4], Ch. 4,3). Summarizing we obtain the following result with

$$c_s = \frac{n!}{(n-m)!} \binom{n+m+s}{n} \binom{n-m}{s} .$$

Sectoral hyperboloid functions ($m \neq 0$)

Generators: $g(z) = \dfrac{z^{n-m}}{2(n-m)!}$ and $g(z) = \dfrac{iz^{n-m}}{2(n-m)!}$,

$$(13)\qquad w_{nm} = \begin{cases} \dfrac{z^m + \bar{z}^m}{2(1-z\bar{z})^m} \sum\limits_{s=0}^{n-m} c_s \left(\dfrac{z\bar{z}}{1-z\bar{z}}\right)^s \\ \\ \dfrac{z^m - \bar{z}^m}{2i(1-z\bar{z})^m} \sum\limits_{s=0}^{n-m} c_s \left(\dfrac{z\bar{z}}{1-z\bar{z}}\right)^s \end{cases} \qquad m = 1, 2, \ldots, n.$$

<u>Nonvanishing hyperboloid function ($m = 0$):</u>

Generator: $g(z) = \dfrac{z^n}{2n!}$,

$$(14)\qquad w_{no} = \sum_{s=0}^{n} c_s \left(\frac{z\bar{z}}{1-z\bar{z}}\right)^s > 0.$$

2) A representation of the surface harmonics of degree n in p dimensions

Let y_k, $k = 1, \ldots, p$, be Cartesian coordinates in the Euclidian space E^p. We set

$$y = (y_1, \ldots, y_p), \quad |y| = \rho \quad \text{and} \quad y_o = \frac{1}{\rho} y$$

and denote by

$$\Delta_p = \sum_{k=1}^{p} \frac{\partial^2}{\partial y_k^2}$$

the Laplace operator. Let $H_n^p(y)$ be a homogeneous polynomial of degree n which satisfies the differential equation

$$\Delta_p H_n^p = 0.$$

Then,

$$H_n^p(y_o) = \frac{1}{\rho^n} H_n^p(y)$$

is called a surface harmonic of degree n in p dimensions. The number N_n^p of linearly independent functions $H_n^p(y)$ is given by

$$N_n^p = \sum_{s=0}^{n} N_s^{p-1} \tag{15}$$

(cf. e.g.[89]). By the stereographic projection the unit hypersphere S^{p-1}: $|y| = 1$ is mapped onto the closed hyperplane E^{p-1}. If x_k, $k = 1, \ldots, p-1$, denote Cartesian coordinates in E^{p-1}, by $x = (x_1, \ldots, x_{p-1})$ and $r = |x|$ it follows

$$y = \frac{\rho}{1+r^2}\begin{pmatrix} 2x \\ r^2-1 \end{pmatrix}.$$

If

$$\Delta_p^* = \frac{1}{\sqrt{g}} \frac{\partial}{\partial x^i} \sqrt{g}\, g^{ik} \frac{\partial}{\partial x^k}$$

denotes the Beltrami operator on S^{p-1}, by means of the tensor calculus we obtain

$$\Delta_p = \frac{\partial^2}{\partial\rho^2} + \frac{p-1}{\rho}\frac{\partial}{\partial\rho} + \frac{1}{\rho^2}\Delta_p^*$$

(cf.[89]). Using

$$\Delta_{p-1} = \sum_{k=1}^{p-1} \frac{\partial^2}{\partial x_k^2}, \quad D_{p-1} = \sum_{k=1}^{p-1} x_k \frac{\partial}{\partial x_k} \quad \text{and } \omega = 1+r^2,$$

we get the differential equation of the surface harmonics $H_n^p(y_o)$

(16) $$\omega^2\Delta_{p-1} v-2(p-3)\omega D_{p-1}v+4n(n-2+p)v = 0, \quad n \in \mathbb{N}_o, \quad p \geq 3,$$

which is transformed by

$$v = \omega^{\frac{p-3}{2}} w$$

into

(17) $$\omega^2\Delta_{p-1}w + 4\left(n+\frac{p-3}{2}\right)\left(n+\frac{p-1}{2}\right) w = 0.$$

In the case $p = 2m+1$, $m \in \mathbb{N}$, we get

(18) $$\omega^2\Delta_{2m}w + 4(n+m-1)(n+m)w = 0.$$

As in the case of the differential equation (I,211) we set

(19) $$w = \sum_{k=0}^{n+m-1} \frac{h_k}{\omega^{n+m-1-k}}, \quad \Delta_{2m}h_k = 0,$$

in order to get solutions of (18). By inserting this into (18) it follows that (19) represents a solution if

$$h = \frac{(n+m-1)!}{(2n+2m-2)!} h_o, \quad h_o(x) \text{ bel.},$$

$$h_k = A_k(D_{2m}-n)_k h, \quad A_k = \frac{[2(n+m-1)-k]!}{k!(n+m-1-k)!}, \quad k = 1, \ldots, n+m-1.$$

Proceeding from this result, we try to determine solutions of (16) which are defined in the closed hyperplane E^{2m}. First of all the summands with $k > n$ are to vanish. This is the case if we use the generators $h = h_j(x)$ with

$$D_{2m} h_j = j h_j , \quad j = 0, 1, \ldots, n, \tag{20}$$

to say the generators are homogeneous harmonic polynomials $h_j = H_j^{2m}(x)$ of degree $j = 0, 1, \ldots, n$. Thus, we get by

$$v_j = \sum_{k=0}^{n-j} A_k \frac{(D_{2m} - n)_k H_j^{2m}(x)}{\omega^{n-k}}$$

respectively

$$v_j = H_j^{2m} \sum_{s=j}^{n} \frac{[2(n+m-1)+s-n]!}{(n-s)!(m-1+s)!} \frac{(j-n)_{n-s}}{\omega^s} \tag{21}$$

solutions of (16) for $p = 2m+1$. However, by (21) we see that the condition (20) is also sufficient. Considering (15), by (21) we obtain the set of all surface harmonics in $p = 2m+1$ dimensions. Summarizing we get the following result (cf.[13]).

Theorem 1

Let $H_j^{2m}(x)$, $j = 0, 1, \ldots, n$, be the most general homogeneous harmonic polynomial in 2m variables of degree j.
Then, by

$$v = \sum_{j=0}^{n} H_j^{2m}(x) Q_{n,j}^{2m}(\omega), \tag{22}$$

with

$$Q_{n,j}^{2m}(\omega) = \sum_{s=j}^{m} \frac{[2(n+m-1)+s-n]!}{(n-s)!(m-1+s)!} \frac{(j-n)_{n-s}}{\omega^s} ,$$

we get a representation of the surface harmonics of degree n in $p=2m+1$ dimensions.

The above procedure is not applicable if p is an even number. However, the result in Theorem 1 suggests that we set

$$v = \sum_{j=0}^{n} v_j$$

with

$$v_j = H_j^{p-1}(x)\varphi\left(\frac{1}{\omega}\right)$$

and

$$\triangle_{p-1} H_j^{p-1}(x) = 0, \quad D_{p-1} H_j^{p-1}(x) = j H_j^{p-1}(x),$$

$$x = (x_1, \ldots, x_{p-1}),$$

in general for $p \geq 3$. By inserting into (16) it follows (cf. [13])

Theorem 2

Let $H_j^{p-1}(x)$, $j = 0,1, \ldots, n$, be the most general homogeneous harmonic polynomials in p-1 variables of degree j.
Then, by

$$v = \sum_{j=0}^{n} H_j^{p-1}(x) \frac{1}{\omega^j} C_{n-j}^{\frac{p-2}{2}+j}\left(1-\frac{2}{\omega}\right) \tag{23}$$

we get a representation of the surface harmonics of degree n in p dimensions if C_μ^λ denote the Gegenbauer polynomials [87].

Proceeding from a representation of the solutions of the differential equation

$$\triangle_p u + 4B(r^2)u = 0 \quad {}^{19)}$$

by integral operators derived in [56] and [57], H. Florian and R. Heersink [49] could deduce a representation by means of differential operators in the case of

19) Cf. also [47] and [68].

$$B(r^2) = n(n+1)\,\frac{em^2r^{2(m-1)}}{(1+er^{2m})^2}\,, \qquad e = \pm 1, \quad m \in \mathbb{N}, \quad n \in \mathbb{N}_0.$$

Moreover, using an assertion proved in [13], a representation of the surface harmonics was derived in [49] which is similar to the result in Theorem 1.

3) Pseudo-analytic functions and complex potentials

a) Representation of the solutions of the differential equation $w_{\bar{z}} = c\bar{w}$ with $m^2(\log c)_{z\bar{z}} = c\bar{c}$, $m \in \mathbb{N}$

After L.Bers (cf.[33,34]) the solutions of the differential equation

$$(24) \qquad W_{\bar{z}} = aW + b\bar{W}$$

are called pseudo-analytic functions. Transforming (24) by

$$W = we^{A}$$

with $a = A_{\bar{z}}$, we obtain

$$(25) \qquad w_{\bar{z}} = c\bar{w} \quad \text{with} \quad c = be^{\bar{A}-A}.$$

If the coefficient c in (25) is analytic and satisfies certain conditions, it is possible to derive general representation theorems for the solutions [20] of (25) defined in simply connected domains D by differential operators using the representations of solutions for elliptic equations treated in Chapter I. This differential equation belongs to the class

$$w_{\bar{z}} = \alpha w + \beta\bar{w} + \gamma, \qquad \alpha, \beta, \gamma \text{ analytic in } D,$$

for which I.N. Vekua (cf. [102], Ch.I,15) developed a complete theory, where the solutions are represented by means of certain integral operators. Since the explicit determination of the required resolvents may be difficult, other representations are of special importance which handle easily. If the coefficient c in (25) satisfies the non-linear differential equation

---

[20] In this section a solution is a continuously differentiable function which satisfies the differential equation (25). Since here the coefficient $c(z,\bar{z})$ is a differentiable function, the solutions satisfy also the elliptic differential equation $cw_{z\bar{z}} - c_z w_{\bar{z}} - c^2\bar{c}w = 0$ and, therefore, are analytic.

(26) $$m^2(\log c)_{z\bar{z}} = c\bar{c}, \quad m \in \mathbb{N},$$

it can be represented in the form

$$c = \frac{\varepsilon m \overline{\alpha'} \beta}{(\alpha+\bar{\alpha})\bar{\beta}}, \qquad \varepsilon = \pm 1,$$

(cf. [29], Theorem 2), where $\alpha(z)$ and $\beta(z)$ are holomorphic in D and satisfy the condition

$$(\alpha + \bar{\alpha})\alpha'\beta \neq 0.$$

Without loss of generality we may set

$$\varepsilon = \beta = 1,$$

as can be shown by simple transformations. Therefore, we proceed from the differential equation

(27) $$w_{\bar{z}} = c\bar{w} \quad \text{with} \quad c = \frac{m\overline{\alpha'}}{\alpha+\bar{\alpha}}, \quad m \in \mathbb{N}.$$

We set

$$w = u + iv, \quad u,v \text{ real-valued},$$

and differentiate (27) with respect to z. Then, we obtain the differential equations

(28) $$(\alpha+\bar{\alpha})^2 u_{z\bar{z}} - \alpha'\overline{\alpha'}m(m-1)u = 0, \quad m \in \mathbb{N},$$

(29) $$(\alpha+\bar{\alpha})^2 v_{z\bar{z}} - \alpha'\overline{\alpha'}m(m+1)v = 0, \quad m \in \mathbb{N}.$$

These equations represent special classes of the differential equation (I,67). By Theorem I,9 the real part u and the imaginary part v of w may be represented in the form

(30) $$u = H_{m-1}g + \overline{H_{m-1}g},$$

$$v = H_m h + \overline{H_m h} , \tag{31}$$

where g(z) and h(z) are arbitrary holomorphic functions in D. If we insert w = u + iv with u and v according to (30) and (31) into (27), by 2g = Rf and 2ih = f (cf.[29]) we obtain the following

Theorem 3

a) For every solution w of the differential equation (27)

$$w_{\bar{z}} = \frac{m\overline{\alpha'}}{\alpha+\bar{\alpha}} \bar{w} , \quad m \in \mathbb{N},$$

with

$$\alpha(z) \in H(D) \quad \text{and} \quad (\alpha+\bar{\alpha})\alpha' \neq 0 \text{ in } D,$$

defined in D, there exists a function f(z) ∈ H(D), such that

$$w = Q_m^* f = \sum_{k=0}^{m} \frac{(-1)^{m-k}(2m-1-k)!}{k!(m-k)!(\alpha+\bar{\alpha})^{m-k}} [mR^k f-(m-k)\overline{R^k f}] \tag{32}$$

with $R = \frac{1}{\alpha'} \frac{\partial}{\partial z}$.

b) Conversely, for each function f(z) ∈ H(D) (32) represents a solution of (27) in D.

c) For every given solution w of (27) the function $R^{2m} f$ is uniquely determined by

$$R^{2m} f = \frac{R^m[(\alpha+\bar{\alpha})^m w]}{(\alpha+\bar{\alpha})^m} . \tag{33}$$

In this case the generator f(z) is not uniquely determined. We obtain the most general generator $\tilde{f}(z)$ by

$$\tilde{f}(z) = f(z) + \sum_{\mu=0}^{2m-1} a_\mu \alpha^\mu , \quad a_\mu \in \mathbb{C},$$

with

$$a_\mu - (-1)^\mu \overline{a_\mu} = 0, \quad \mu = 0,1, \ldots, 2m-1.$$

Corollary

Because of (33) for every solution w of (27) which is defined in a (not necessarily simply connected) domain D the function $R^{2m}f$ is uniquely determined in each point of D and represents a globally unique holomorphic function in D.

If we denote the set of the solutions of (27) defined in D by $F_m(D)$ and if we use the differential operators

$$R = \frac{1}{\alpha^\tau}\frac{\partial}{\partial z}, \qquad S = \frac{1}{\overline{\alpha^\tau}}\frac{\partial}{\partial \bar{z}},$$

we obtain the following assertions, as easily can be verified.

Theorem 4

Let $w = Q_m^* f \in F_m(D)$. Then,

a) $i(R-S)w = Q_m^*(iRf) \in F_m(D), \quad m \in \mathbb{N}.$

b) $\left(R + \frac{m+1}{m} S - \frac{2m+1}{\alpha+\bar{\alpha}}\right) w = Q_{m+1}^* f \in F_{m+1}(D), \quad m \in \mathbb{N},$

$$Rw + \frac{m+1}{\alpha+\bar{\alpha}}\bar{w} - \frac{2m+1}{\alpha+\bar{\alpha}} w = Q_{m+1}^* f \in F_{m+1}(D), \quad m \in \mathbb{N}_0.$$

c) $\left(R + \frac{m-1}{m} S + \frac{2m-1}{\alpha+\bar{\alpha}}\right) w = Q_{m-1}^*(R^2 f) \in F_{m-1}(D), \quad m \in \mathbb{N}.$

Proceeding from the holomorphic functions, by Theorem 4,b it is possible successively to get the solutions of (27) and to derive the representation (32). In order to prove equation (33) it is more advantageous to use the representation (32) instead of the relation in Theorem 4,c.

By the Corollary to Theorem 3 we obtain a general expansion theorem for the solutions of (27) in the neighbourhood of isolated singularities by a corresponding procedure as in the case of the differential equation (I,56). In the following we quote this assertion and refer the reader to [29], p. 9-11, for further details.

Theorem 5

Let w be a solution of (27) in

$$\dot{U}(z_o) = \{ z \mid 0 < |z-z_o| < \rho \}$$

with an isolated singularity at z_o. Then, w can be represented in $\dot{U}(z_o)$ by

$$w = \Omega_m^* f$$

with the generator

$$f(z) = f_1(z) + p(\alpha) \log(z-z_o)$$

where $f_1(z)$ is a holomorphic and unique function in $\dot{U}(z_o)$, whereas $p(\alpha)$ represents a polynomial

$$p(\alpha) = \sum_{\mu=0}^{2m-1} b_\mu \alpha^\mu , \quad b_\mu \in \mathbb{C},$$

with

$$b_\mu + (-1)^\mu \overline{b_\mu} = 0.$$

Every solution of (25) satisfies also the elliptic differential equation

$$(34) \qquad w_{z\bar{z}} - \frac{c_z}{c} w_{\bar{z}} - c\bar{c}w = 0$$

if the coefficient c in (25) is a differentiable function (cf.[101], p.140).

Conversely, proceeding from a solution w of (34), we obtain solutions of (25) by

$$w_1 = \frac{1}{2}\left(w + \frac{1}{\bar{c}}\bar{w}_z\right), \quad w_2 = \frac{1}{2i}\left(w - \frac{1}{\bar{c}}\bar{w}_z\right).$$

It follows that the solution of (34) may be represented by

$$w = w_1 + iw_2. \tag{35}$$

After I.N. Vekua (cf. [101], p.141) the solutions of (34) are called complex potentials of the differential equation (25).

Proceeding from the differential equation (27)

$$w_{\bar{z}} = \frac{m\overline{\alpha'}}{\alpha+\bar{\alpha}}\bar{w}$$

the complex potentials are solutions of the differential equation

$$w_{z\bar{z}} + \frac{\alpha'}{\alpha+\bar{\alpha}} w_{\bar{z}} - m^2 \frac{\alpha'\overline{\alpha'}}{(\alpha+\bar{\alpha})^2} w = 0. \tag{36}$$

Here we get a particular case of the equation (I,56). Using Theorem 3,a and (35), we obtain the solutions of (36) defined in D by

$$w = Q_m^* f_1 + iQ_m^* f_2, \quad f_1, f_2 \in H(D).$$

If we set

$$g = m(f_1+if_2) \text{ and } f = f_1 - if_2,$$

it follows

$$w = \sum_{k=0}^{m} \frac{(-1)^{m-k}(2m-1-k)!}{k!(m-k)!(\alpha+\bar{\alpha})^{m-k}} \left[R^k g - (m-k)\overline{R^k f}\right]$$

(cf.(I,57)).

In connection with the representation of pseudo-holomorphic functions of several complex variables A. Koohara [73] obtained the differential equation

$$G_{\bar{z}} = \frac{\bar{K}K_{\bar{z}}}{1-K\bar{K}} G - \frac{\overline{K_z}}{1-K\bar{K}} \bar{G} \tag{37}$$

and asked the following question (cf. [73], p.273): Is there a non-constant function $K(z,\bar{z})$, such that equation (37) is reduced to a differential equation of the form

$$w_{\bar{z}} = c\bar{w} ,$$

where the coefficient c satisfies the conditions (26). In fact, such functions $K(z,\bar{z})$ exist, as will be proved by two examples (cf. [23]).

For instance, setting

$$K = e^{i\mu} \frac{1+\varepsilon(\alpha+\bar{\alpha})^{2m}}{1-\varepsilon(\alpha+\bar{\alpha})^{2m}} \tag{38}$$

with

$$\varepsilon = \pm 1, \qquad \mu \in \mathbb{R}, \quad m \in \mathbb{N}, \quad 1-\varepsilon(\alpha+\bar{\alpha})^{2m} \neq 0,$$

the differential equation (37) is transformed by

$$G = e^{-\frac{i\mu}{2}} \frac{1-\varepsilon(\alpha+\bar{\alpha})^{2m}}{(\alpha+\bar{\alpha})^{m}} w$$

into (27). By using (32) we get the corresponding solution of (37). If we employ especially the functions

$$f = \frac{\alpha^{2m}}{m!} \quad \text{and} \quad f = \frac{i(-1)^m m!}{(2m)!} ,$$

we obtain particular solutions which form a generating pair in L. Bers's sense [33].

Theorem 6

a) For every solution of the differential equation (37) in D with K according to (38) and

$$\alpha(z) \in H(D), \quad (\alpha+\bar{\alpha})\alpha' \neq 0 \text{ in } D,$$

there exist a function $f(z) \in H(D)$, such that

$$(39) \quad G = e^{-\frac{i\mu}{2}} \frac{1-\varepsilon(\alpha+\bar{\alpha})^{2m}}{(\alpha+\bar{\alpha})^{2m}} \sum_{k=0}^{m} \frac{(-1)^{m-k}(2m-1-k)!}{k!(m-k)!}(\alpha+\bar{\alpha})^{k}[mR^{k}f-(m-k)\overline{R^{k}f}].$$

b) Conversely, for each function $f(z) \in H(D)$ (39) represents a solution of (37) in D with K according to (38).

c) The particular solutions

$$G_1 = e^{-\frac{i\mu}{2}}[1-\varepsilon(\alpha+\bar{\alpha})^{2m}], \quad G_2 = ie^{-\frac{i\mu}{2}} \frac{1-\varepsilon(\alpha+\bar{\alpha})^{2m}}{(\alpha+\bar{\alpha})^{2m}}$$

form a generating pair in L. Bers's sense.

Using, for example,

$$(40) \quad k = \frac{\gamma^{\lambda+1}}{\bar{\gamma}^{\lambda}}, \quad \lambda \in \mathbb{Z}, \quad \gamma(z) \in H(D), \quad (1-\gamma\bar{\gamma})\gamma\gamma' \neq 0 \text{ in } D,$$

the differential equation (37) is transformed by

$$G = (1-\gamma\bar{\gamma})^{\lambda} v$$

into

$$(41) \quad v_{\bar{z}} = c\bar{v} \quad \text{with} \quad c = -\frac{(\lambda+1)\bar{\gamma}^{\lambda}\overline{\gamma'}}{\gamma^{\lambda}(1-\gamma\bar{\gamma})}.$$

In the case $\lambda = -1$ we obtain

$$K = \overline{\gamma(z)}, \quad v_{\bar{z}} = 0, \quad G = \frac{f(z)}{1-\gamma\bar{\gamma}}, \quad f(z) \in H(D).$$

In case $\lambda \neq -1$ by $\gamma = \frac{\alpha-1}{\alpha+1}$ the coefficient c gets the form

$$c = \frac{-(\lambda+1)\overline{\alpha'}}{\alpha+\bar{\alpha}} \frac{(\alpha-1)^{\lambda+1}}{(\alpha+1)^{\lambda}} \frac{(\bar{\alpha}+1)^{\lambda}}{(\bar{\alpha}-1)^{\lambda+1}},$$

and the differential equation (41) is transformed into (27) by

$$v = \beta w$$

if we use

$$\beta = \frac{2i}{\gamma^{\lambda}(\gamma-1)} \quad \text{for} \quad m = \lambda+1 \quad \text{and } \lambda \in \mathbb{N}_0$$

and

$$\beta = \frac{2}{\gamma^{\lambda}(\gamma-1)} \quad \text{for} \quad m = -(\lambda+1) \quad \text{and} \quad -\lambda \in \mathbb{N}, \quad \lambda \leq -2.$$

Theorem 7

a) For every solution of (37) in D with K according to (40) and

$$\gamma(z) \in H(D), \quad (1-\gamma\bar{\gamma})\gamma\gamma' \neq 0 \text{ in } D,$$

there exists a function $f(z) \in H(D)$, such that

$$(42) \qquad G = \begin{cases} i\left(\frac{1-\gamma\bar{\gamma}}{\gamma}\right)^{\lambda} P_{\lambda+1}f & \text{for} \quad \lambda \in \mathbb{N}_0, \\ \frac{f(z)}{1-\gamma\bar{\gamma}} & \text{for} \quad \lambda = -1, \\ \left(\frac{\gamma}{1-\gamma\bar{\gamma}}\right)^{|\lambda|} P_{|\lambda|-1}f & \text{for} \quad -\lambda \in \mathbb{N}, \quad \lambda \leq 2, \end{cases}$$

with

$$P_m f = \sum_{k=0}^{m} \frac{(-1)^{m-k}(2m-1-k)!(\gamma-1)^{m-k-1}(\bar{\gamma}-1)^{m-k}}{k!(m-k)!(1-\gamma\bar{\gamma})^{m-k}}[mS^k f-(m-k)\overline{S^k f}],$$

$$S = \frac{(\gamma-1)^2}{\gamma'}\frac{\partial}{\partial z}, \quad m \in \mathbb{N}.$$

b) Conversely, for each function $f(z) \in H(D)$ (42) represents a solution of (37) in D with K according to (40).

c) The particular solutions ($\lambda \in \mathbb{Z}$)

$$G_1 = \frac{(\gamma-1)^{\lambda}(\bar{\gamma}-1)^{\lambda+1}}{\gamma^{\lambda}(1-\gamma\bar{\gamma})}, \quad G_2 = \frac{i(1-\gamma\bar{\gamma})^{2\lambda+1}}{\gamma^{\lambda}(\gamma-1)^{\lambda+2}(\bar{\gamma}-1)^{\lambda+1}}$$

form a generating pair in L.Bers's sense.

As is wellknown, there are pseudo-analytic functions which attain the maximum of their absolute value in the interior of the domain of definition. However, G. Jank and K.-J. Wirths [71] proved that there are certain classes of pseudo-analytic functions for which a generalized "sharp" maximum principle is valid.

Theorem (G.Jank and K.-J. Wirths)

Let $g(z)$ be a holomorphic and nonvanishing function in D. Let γ be a twice continuously differentiable real-valued function in D and γ^{-2} subharmonic in D. If w is a twice continuously differentiable solution of the differential equation

$$w_{\bar{z}} = \frac{\gamma_{\bar{z}}}{\gamma} \frac{g}{\bar{g}} \bar{w} .$$

in D with

$$\overline{\lim_{z \to \zeta}} \left|\frac{w}{g\gamma}\right| \leq 1$$

for each $\zeta \in \partial D$, then

$$|w| \leq |g\gamma| \text{ in } D.$$

If

$$|w(z_o)| = |g(z_o)\gamma(z_o)|,$$

for some $z_o \in D$, then

$$|w| = |g\gamma| \text{ in } D.$$

Using, for example, $\gamma = (\alpha+\bar{\alpha})^{\varepsilon m}$, $\varepsilon = \pm 1$, $m \in \mathbb{N}$, $\alpha(z) \in H(D)$, $(\alpha+\bar{\alpha})\alpha' \neq 0$ in D, and $g = \sqrt{-\varepsilon}$, then γ^{-2} is subharmonic, and we get

$$\frac{\gamma_{\bar{z}}}{\gamma}\frac{g}{\bar{g}} = \frac{m\overline{\alpha'}}{\alpha+\bar{\alpha}} .$$

Using the representation (32), the possibility offers to construct non-trivial pseudo-analytic functions for which the above-mentioned "sharp" maximum principle is valid.

If the coefficient c in (25), instead of (26), satisfies the more general differential equation

$$m^2(\log c)_{z\bar{z}} + \varepsilon c\bar{c} = 0, \quad m > 0, \quad \varepsilon = \pm 1,$$

a representation of the corresponding pseudo-analytic functions by differential operators is not possible in general. However, in this case we can explicitly determine the Vekua resolvents required for the representation by integral operators. In these resolvents hypergeometric series appear (cf. [22]) which reduce to polynomials in the case $\varepsilon = -1$, $m \in \mathbb{N}$. Here, the representation may be converted to a form free of integrals, and the result of Theorem 3 can be derived in another way. In the case $\varepsilon = -1$, $m = 1/2$ we obtain a new class of solutions of the Ernst equation (cf. [43-45,35,36]). This non-linear second-order partial differential equation appears in general relativity in connection with the determination of the gravitational field of a uniformly rotating axially symmetric source. In context with the representation of the mentioned solutions the reader is referred to [22].

A generalization of the representation of the pseudo-analytic functions considered in Theorem 3 was investigated by G. Jank and St. Ruscheweyh in [70]. We summarize the results derived in that paper in the following

Theorem 8 (G. Jank and St. Ruscheweyh)

Let $\psi(\tau)$, $\tau = z-\bar{z}$, be a real-analytic and real-valued function in a domain D. Let D* be a simply connected domain in D which contains no point of the imaginary axis. d denotes the differential operator

$$d = \frac{\partial}{\partial z} + \frac{\partial}{\partial \bar{z}} .$$

a) The set of the solutions of the differential equation

$$(43) \qquad w_{\bar{z}} = \left(\frac{n}{\sigma} + i\psi(\tau) \right) \bar{w}, \quad n \in \mathbb{N},\ \sigma = z+\bar{z},$$

defined in D* coincides with the set of functions

$$(44) \qquad w = i \sum_{s=0}^{n} \frac{A_s^n}{(2n-s)\sigma^{n-s}} [nd^s h - sd^{s-1}(h_{\bar{z}} + i\psi(\tau)h)]$$

where h represents an arbitrary real-valued solution of the differential equation

$$(45) \qquad h_{z\bar{z}} + [i\psi'(\tau) - \psi^2(\tau)]h = 0$$

and

$$A_s^n = \frac{(-1)^{n-s}(2n-s)!}{s!(n-s)!} .$$

b) In the case

$$i\,\psi(\tau) = -\frac{m}{\tau}, \quad m \in \mathbb{N},$$

also the solutions of the differential equation (45) may be represented by differential operators (cf. [66,69]) and we obtain instead of (44)

$$w = \sum_{\nu=0}^{n} \sum_{\mu=0}^{m} \frac{1}{\sigma^{n-\nu}\tau^{m-\mu}} \{[A_\nu^n A_\mu^m + A_{\nu-1}^n A_{\mu-1}^m] f^{(\mu+\nu)}(z) -$$

$$- (-1)^{m-\mu}[A_\nu^n A_\mu^m - A_{\nu-1}^{n-1} A_{\mu-1}^{m-1}] \overline{f^{(\mu+\nu)}(z)}\}, \qquad f(z) \in H(D).$$

c) The set of the solutions of the differential equation of the complex potentials

$$w_{z\bar{z}} - \frac{c_z}{c} w_{\bar{z}} - c\bar{c}w = 0, \qquad c = \frac{n}{\sigma} + i\psi(\tau)$$

defined in D coincides with the set of functions (44) if now h denotes an arbitrary complex-valued solution of (45).

b) Representation of pseudo-analytic functions by means of solutions of the generalized Darboux equation

For further differential equations of the form

$$w_{\bar{z}} = cw$$

we get representations of the solutions by differential operators if we suppose that the coefficient c satisfies the condition

$$(46) \qquad c_z \text{ real-valued},$$

and if we apply the results proved for the solutions of the generalized Darboux equation (cf. Chapter I,5). Setting

$$w = u + iv,$$

the real part u and the imaginary part v of a solution w of (25) satisfy the differential equations

$$(47) \qquad u_{z\bar{z}} - (c\bar{c} + c_z)u = 0,$$

$$(48) \qquad v_{z\bar{z}} - (c\bar{c} - c_z)v = 0$$

if c satisfies the condition (46). Using, for instance,

$$(49) \qquad c = \frac{-n}{z+\bar{z}} + \frac{m}{z-\bar{z}}, \qquad m,n \in \mathbb{N},$$

c_z satisfies (46), and we obtain for u and v the differential equations

$$u_{z\bar{z}} + \left[\frac{-n(n+1)}{(z+\bar{z})^2} + \frac{m(m+1)}{(z-\bar{z})^2}\right]u = 0,$$

$$v_{z\bar{z}} + \left[\frac{-(n-1)n}{(z+\bar{z})^2} + \frac{(m-1)m}{(z-\bar{z})^2}\right]v = 0.$$

We get the solutions of these differential equations by Theorem I,23. In this case the differential equation of the complex potentials runs

$$w_{z\bar{z}} + \frac{n(z-\bar{z})^2-m(z+\bar{z})^2}{(z^2-\bar{z}^2([n(z-\bar{z})-m(z+\bar{z})]}\, w_{\bar{z}} - \left[\frac{n^2}{(z+\bar{z})^2} - \frac{m^2}{(z-\bar{z})^2}\right] w = 0.$$

The differential equation (25) with c according to (49) represents a particular case of the differential equation (43) treated in Theorem 8.

Using

$$c = \frac{m}{z+\bar{z}} + \frac{n}{z-\bar{z}} + \varepsilon\,\frac{(n-m)\bar{z}}{1+\varepsilon z\bar{z}}\,, \quad \varepsilon = \pm 1, \quad n,m \in \mathbb{N},$$

c_z satisfies again the condition (46), and we get for u and v the generalized Darboux equations

$$u_{z\bar{z}} + \left[\frac{-(m-1)m}{(z+\bar{z})^2} + \frac{n(n+1)}{(z-\bar{z})^2} + \varepsilon\,\frac{(n-m)(n-m-1)}{(1+\varepsilon z\bar{z})^2}\right] u = 0,$$

$$v_{z\bar{z}} + \left[\frac{-m(m+1)}{(z+\bar{z})^2} + \frac{(n-1)n}{(z-\bar{z})^2} + \varepsilon\,\frac{(n-m)(n-m+1)}{(1+\varepsilon z\bar{z})^2}\right] v = 0.$$

We obtain also the solutions of these differential equations by Theorem I,23. Here, the complex potentials satisfy the differential equation

$$w_{z\bar{z}} + \frac{\dfrac{m}{(z+\bar{z})^2} + \dfrac{n}{(z-\bar{z})^2} - \varepsilon\,\dfrac{n-m}{(1+\varepsilon z\bar{z})^2}}{\dfrac{m}{z+\bar{z}} + \dfrac{n}{z-\bar{z}} + \varepsilon\dfrac{(n-m)z}{1+\varepsilon z\bar{z}}}\, w_{\bar{z}} -$$

$$-\left[\frac{m^2}{(z+\bar{z})^2} - \frac{n^2}{(z-\bar{z})^2} - \frac{\varepsilon(n-m)^2}{(1+\varepsilon z\bar{z})^2}\right] w = 0.$$

c) Representation of pseudo-analytic functions by integro-differential-operators

In the following we consider the differential equation

(50) $$w_{\bar{z}} = \frac{\gamma_{\bar{z}}}{\gamma}\,\bar{w}\,,$$

where γ denotes a real-valued nonvanishing solution of the differential equation (I,156), defined in a simply connected domain D. Then, γ may be represented by

(51) $$\gamma = H\varphi + H\bar{\varphi}\,, \quad \varphi(z) \in H(D).$$

Setting

$$w = u + iv, \quad u,v \text{ real-valued},$$

u and v satisfy the differential equations (I,156) respectively (I, 158). Thus, the results of Theorem I,9 and Theorem I,22 are applicable, and the solutions of (50) have necessarily the form

$$w = Hg + \overline{Hg} + i\left[Hf + \overline{Hf} - \frac{K_{\varphi\varphi}f + \overline{K_{\varphi\varphi}f}}{\gamma}\right].$$

If we substitute into (50), it follows an additional condition for the generators, and we get the following result (cf.[19]).

<u>Theorem 9</u>

Let γ be a real-valued nonvanishing solution of (I,156) in D with the representation (51).

a) For every solution of the differential equation (50)

$$w_{\bar{z}} = \frac{\gamma_{\bar{z}}}{\gamma}\,\bar{w}$$

in D there exists a function $f(z) \in H(D)$, such that

(52) $$w = 2\,iHf - \frac{i}{\gamma}(K_{\varphi\varphi}f + \overline{K_{\varphi\varphi}f}).$$

b) Conversely, for each function $f(z) \in H(D)$ (52) represents a solution of (50) in D.

c) For every given solution w of (50) in D the function $Hf-\overline{Hf}$ is uniquely determined by

$$2\, i(Hf-\overline{Hf})S\gamma = S(\gamma w).$$

In this case the generator f(z) is not uniquely determined. We obtain the most general generator $\tilde{f}(z)$ by $\tilde{f} = f+f_o$ with

$$f_o = \sum_{\mu=0}^{2n} a_\mu \alpha^\mu , \quad a_\mu \in \mathbb{C}, \quad a_\mu - (-1)^\mu \overline{a_\mu} = 0,$$

$$K_{\varphi\varphi} f_o + \overline{K_{\varphi\varphi} f_o} = 2\gamma H f_o .$$

d) The particular solutions

$$w_1 = C_1\gamma, \quad w_2 = iC_2\gamma^{-1}, \quad C_1, C_2 \in \mathbb{R},$$

represent the only real respectively imaginary solutions of (50) in D and yield by $C_1 = C_2 = 1$ a generating pair in L. Bers's sense for the solutions of (50) in D.

In the case of special real-valued solutions γ of (I,156) we may obtain representations free on integrals for the corresponding pseudoanalytic functions; for example in the case

$$\gamma = D_1(\alpha+\bar{\alpha})^{n+1} + D_2(\alpha+\bar{\alpha})^{-n}, \quad D_1, D_2 \in \mathbb{R},$$

(cf. [19]).

4) A generalized Tricomi equation

a) Representation of the solutions in the elliptic respectively hyperbolic half-plane

The theory of transonic flow is closely related to partial differential equations of mixed type. Thus, for example, the hodograph method for plane flows of compressible fluids leads to the differential equation

$$k(\vartheta)\,\psi_{\xi\xi} + \psi_{\vartheta\vartheta} = 0, \tag{53}$$

where $\psi(\xi,\vartheta)$ denotes the stream function, whereas the coefficient $k(\vartheta)$ represents a monotone function with $\vartheta k(\vartheta) > 0$ for $\vartheta \neq 0$. By $k(\vartheta) = \vartheta$ we get the well known Tricomi equation

$$\vartheta\,\psi_{\xi\xi} + \psi_{\vartheta\vartheta} = 0.$$

By a suitable transformation

$$\vartheta = \vartheta(\eta), \quad \Psi(\xi,\eta) = \psi(\xi,\vartheta)$$

(53) may be converted into the canonical form

$$\eta\,\Psi_{\xi\xi} + \Psi_{\eta\eta} + a(\eta)\Psi_{\eta} = 0.$$

If the coefficient $a(\eta)$ has the special form

$$a(\eta) = \frac{6m-1}{2\eta}, \quad m \in \mathbb{Z},$$

we obtain the differential equation

$$\eta\,\Psi_{\xi\xi} + \Psi_{\eta\eta} + \frac{6m-1}{2\eta}\Psi_{\eta} = 0, \quad m \in \mathbb{Z}, \tag{54}$$

for which we may get solutions by differential operators applying the results of Theorem I,9 (cf. [15]).

By the transformation

$$(55) \qquad x = \xi\ , \quad y = \tfrac{2}{3}\eta^{\frac{3}{2}}, \quad \Phi(x,y) = \Psi(\xi,\eta)$$

in the elliptic half-plane ($\eta>0$) we get the differential equation

$$(56) \qquad \Phi_{xx} + \Phi_{yy} + \frac{2m}{y}\Phi_y = 0.$$

By

$$(57) \qquad x = \xi\ , \quad t = \tfrac{2}{3}(-\eta)^{\frac{3}{2}}, \quad X(x,t) = \Psi(\xi,\eta)$$

in the hyperbolic half-plane ($\eta<0$) we get the equation

$$(58) \qquad X_{xx} - X_{tt} - \frac{2m}{t}X_t = 0,$$ [21)]

which may be transformed into the Euler-Poisson-Darboux equation. In the following D denotes a simply connected domain of the upper ξ,η-half-plane. Let D* be the image of D under the transformation (55). By the transition to a complex variable and by the transformation

$$(59) \qquad z = x+iy, \quad v(z,\bar{z}) = i^{-m}(z-\bar{z})^m\Phi(x,y),$$

we obtain the differential equation

$$(60) \qquad (z-\bar{z})^2 v_{z\bar{z}} + (m-1)mv = 0, \quad m \in \mathbb{Z},$$

instead of (56). Setting

$$(61) \qquad n = m-1 \quad \text{for} \quad m \in \mathbb{N}$$

and

$$(62) \qquad n = -m \quad \text{for} \quad -m \in \mathbb{N}_0,$$

we get

---

21) For the equations (56) and (58) we refer the reader also to [46, 52,54,55,63,64,110,111,113].

$$(z-\bar{z})^2 v_{z\bar{z}} + n(n+1)v = 0, \quad n \in \mathbb{N}_0, \tag{63}$$

that is we get a particular case of the differential equation (I,67). Considering (61) and (62) we obtain the real-valued solutions of (60) in the form

$$v = \frac{(z-\bar{z})^{n+1}}{i^m} T^n \frac{f(z)+(-1)^{n+m}\overline{f(z)}}{z-\bar{z}}, \quad T = \frac{\partial^2}{\partial z \partial \bar{z}}, \tag{64}$$

where $f(z)$ denotes an arbitrary holomorphic function in the considered domain. Thus, for the solutions of the differential equation (54) we obtain (cf. [15]) the following

Theorem 10

a) For every solution of the differential equation (54)

$$\eta \Psi_{\xi\xi} + \Psi_{\eta\eta} + \frac{6m-1}{2\eta} \Psi_{\eta} = 0, \quad m \in \mathbb{Z},$$

defined in D, there exists a function $f(z) \in H(D^*)$, such that

$$\Psi(\xi,\eta) = \left[(z-\bar{z})^{n+1-m} T^n \frac{f(z)+(-1)^{n+m}\overline{f(z)}}{z-\bar{z}} \right]_{z = \xi + \frac{2i}{3}\eta^{\frac{3}{2}}}. \tag{65}$$

Here and in the following assertions we set

$$n = m-1 \quad \text{for} \quad m \in \mathbb{N}, \quad n = -m \quad \text{for} \quad -m \in \mathbb{N}_0. \tag{66}$$

b) Conversely, for each function $f(z) \in H(D^*)$ (65) represents a solution of (54) in D.

c) For every given solution $\Psi(\xi,\eta)$ of (54) in D the function $f^{(2n+1)}(z)$ is uniquely determined by

$$f^{(2n+1)}(x+iy) = \frac{i^m}{2^{n+1-m} n! \, y^{2n+2}} K^{n+1} \left[y^m \Psi\left(x, \left(\frac{3y}{2}\right)^{\frac{2}{3}}\right) \right]$$

where K denotes the differential operator

$$K = y^2\left(\frac{\partial}{\partial x} - i\,\frac{\partial}{\partial y}\right).$$

In this case the generator $f(z)$ is not uniquely determined. We obtain the most general generator $\tilde{f}(z)$ by

$$\tilde{f}(z) = f(z) + \sum_{\mu=0}^{2n} a_\mu z^\mu \quad \text{with} \quad a_\mu + (-1)^n\overline{a_\mu} = 0.$$

By virtue of (66) we have in (65)

$$f(z) + (-1)^{n+m}\overline{f(z)} = \begin{cases} f(z) - \overline{f(z)} & \text{for} \quad m \in \mathbb{N}, \\ f(z) + \overline{f(z)} & \text{for} \quad -m \in \mathbb{N}_0. \end{cases}$$

To say in any case a harmonic function appears as generator. Thus, by suitable normalization we obtain the representation

$$\Phi(x,y) = \frac{y^{n+1-m}}{(-2)^n n!}\,\Delta^n\,\frac{u(x,y)}{y} \tag{67}$$

for the solutions of (56) in D^*, where $u(x,y)$ is an arbitrary harmonic function in D^* and Δ denotes the Laplace operator

$$\Delta = \frac{\partial^2}{\partial x^2} + \frac{\partial^2}{\partial y^2}$$

whereas n is to insert according to (66). Moreover, if we imploy the operators

$$D_y = \frac{\partial}{\partial y} \quad \text{and} \quad d_y = \frac{1}{y}\,D_y\,,$$

we get

$$\Phi = \sum_{k=0}^{n} \frac{(2n-k)!}{(-2)^{n-k}k!(n-k)!}\,\frac{D_y^k u}{y^{n+m-k}} = y^{n+1-m} d_y^n\,\frac{u}{y}$$

instead of (67).

Theorem 11

a) For every solution $\Psi(\xi,\eta)$ of the differential equation (54)

$$\eta \Psi_{\xi\xi} + \Psi_{\eta\eta} + \frac{6m-1}{2}\Psi_{\eta} = 0, \quad m \in \mathbb{N},$$

defined in D, there exists a harmonic function u(x,y) in D*, such that

$$\Psi(\xi,\eta) = \left\{ y^{n+1-m} d_y^n \frac{u}{y} \right\}_{x=\xi,\ y=\frac{2}{3}\eta^{\frac{3}{2}}} \tag{68}$$

with n according to (66).

b) Conversely, for each harmonic function u(x,y) in D* (68) represents a solution of (54) in D.

c) For every given solution $\Psi(\xi,\eta)$ of (54) in D the function D_y^{2n} u(x,y) is uniquely determined by

$$D_y^{2n} u = d_y^n \left[y^{n+m} \Psi\left(x, \left(\frac{3y}{2}\right)^{\frac{2}{3}}\right)\right].$$

In this case, apart from m = 0 and m = 1, the generator u(x,y) is not uniquely determined. We obtain the most general generator $\tilde{u}(x,y)$ by

$$\tilde{u}(x,y) = u(x,y) + \sum_{\mu=0}^{n-1} g_\mu(x) y^{2\mu+1}$$

with

$$g_\mu(x) = (-1)^\mu \sum_{s=0}^{2(n-\mu)-1} d_s \binom{2n-s}{2\mu+1} x^{2(n-\mu)-1-s},$$

$$d_s \in \mathbb{R}, \quad \mu = 0,1,\ldots,n-1.$$

Transforming (58) by

$$\psi(r,s) = X(x,t), \quad r = x+t, \quad s = x-t,$$

it follows the Euler-Poisson-Darboux equation

$$(r-s)\psi_{rs} - m(\psi_r - \psi_s) = 0$$

which has the solution (cf. [40], Ch.III)

$$\psi(r,s) = (r-s)^{n+1-m} \left[\frac{\partial^2}{\partial r \partial s}\right]^n \frac{\alpha(r)+\beta(s)}{r-s}$$

with n according to (66). Thus, we get the solutions of (58) by

$$X(x,t) = t^{n+1-m} H^n \frac{\gamma(x,t)}{t} \tag{69}$$

with

$$H = \frac{\partial^2}{\partial x^2} - \frac{\partial^2}{\partial t^2},$$

if $\gamma(x,t)$ denotes an arbitrary solution of the differential equation

$$H\gamma = 0.$$

Similarly to the solutions of (56) we may represent the solutions (69), it follows

$$X(x,t) = t^{n+1-m} \sum_{k=0}^{n} (-2)^k (2n-k)! \binom{n}{k} \frac{D_t^k \gamma}{t^{2n+1-k}} . \tag{70}$$

b) Fundamental solutions in the large

For the generalized Tricomi equation (54) we may determine fundamental solutions in the large, applying the above representations of solutions and the assertions about the solutions of the differential equation (cf. (I,92))

$$(1-\zeta\bar{\zeta})^2 w_{\zeta\bar{\zeta}} - n(n+1)w = 0, \quad n \in \mathbb{N}, \tag{71}$$

which are defined in $|\zeta| < 1$ up to isolated singularities and bounded for $|\zeta| \to 1$ (cf. [99]).
Let B^n be the set of the above-characterized solutions of (71). Let H be the set of functions

$$h(\zeta) = h^*(\zeta) + \sum_{k=1}^{s} s_k(\zeta) \log \frac{\zeta - \zeta_k}{1-\bar{\zeta}_k \zeta} , \tag{72}$$

with:

(i) Up to isolated singularities $h^*(\zeta)$ is a unique holomorphic function on the Riemann number sphere which has no singularities on $|\zeta| = 1$.

(ii) $|\zeta_k| < 1$ for $k = 1, \ldots, s$.

(iii) The functions $S_k(\zeta)$, $k = 1, \ldots, s$, denote polynomials in ζ of degree $\leq n$.

Theorem 12 (St. Ruscheweyh)

a) If $w \in B^n$, then

$$(73) \qquad |w(\zeta,\bar{\zeta})| = O[(1-\zeta\bar{\zeta})^{n+1}] \quad \text{for} \quad |\zeta| \to 1.$$

b) For every real-valued solution $w \in B^n$ there exists a generator $h(\zeta) \in H$, satisfying the condition

$$(74) \qquad h(\zeta) = -\zeta^{2n}\overline{h\left(\frac{1}{\bar{\zeta}}\right)},$$

such that

$$(75) \qquad w(\zeta,\bar{\zeta}) = E_n h + \overline{E_n h}$$

in $|\zeta| < 1$ with

$$(76) \qquad E_n = \sum_{k=0}^{n} \frac{(2n-k)!}{k!(n-k)!} \left[\frac{\bar{\zeta}}{1-\zeta\bar{\zeta}}\right]^{n-k} \frac{d^k}{dz^k}.$$

c) Conversely, for each function $h(\zeta) \in H$, satisfying the condition (74), (75) represents a real-valued solution of the set B^n.

d) If $w \in B^n$, then we may real-analytically continue w outside the unit circle, and for this continuation it follows

$$w(\zeta,\bar{\zeta}) = (-1)^{n+1} w\left(\frac{1}{\zeta}, \frac{1}{\bar{\zeta}}\right), \quad |\zeta| > 1.$$

Transforming the differential equation (63) by

$$\zeta = \frac{i-z}{i+z}\,, \qquad w(\zeta,\bar{\zeta}) = v(z,\bar{z}),$$

we get (71). If we proceed from a solution (75), it follows by Theorem I,9

$$v = \frac{(z-\bar{z})^{n+1}}{n!}\, T^n\, \frac{f(z)+\overline{g(z)}}{z-\bar{z}} \tag{77}$$

with

$$f(z) = \frac{(i+z)^{2n}}{(-2i)^n}\, h\left(\frac{i-z}{i+z}\right),$$

$$g(z) = \frac{(i+z)^{2n}}{(2i)^n}\, h\left(\frac{i-z}{i+z}\right) = (-1)^n f(z). \tag{78}$$

In place of (72) we have

$$f(z) = f^*(z) + \sum_{k=1}^{s} S_k^*(z)\, \log \frac{z_k - z}{\overline{z_k} - z}\,, \tag{79}$$

with

(i) Up to isolated singularities $f^*(z)$ is a unique holomorphic function on the Riemann number sphere which has no singularities on the real axis.

(ii) Im $z_k > 0$ for $k = 1, \ldots, s$.

(iii) The functions $S_k^*(z)$, $k = 1, \ldots, s$, denote polynomials in z of degree $\leq 2n$.

Moreover, on account of (74) $f(z)$ has to satisfy the condition

$$f(z) = (-1)^{n+1}\, \overline{f(\bar{z})}\,. \tag{80}$$

Now we use the generator

$$f(z) = (-i)^n 2^{n+m} n!\, \log \frac{z-ib}{z+ib}\,, \quad b > 0. \tag{81}$$

This function belongs to the class characterized by (79), it satisfies the condition (80) and by (77) and (78) it yields a real-valued solution

$$v(z,\bar{z}) = \frac{(z-\bar{z})^{n+1}}{n!} T^n \frac{f(z)+f(\bar{z})}{z-\bar{z}}$$

of (63) which is defined in the upper half-plane up to $z_o = ib$, $b > 0$, and which tends to zero of order $n+1$ as z tends to the real axis. Transforming by (59) we obtain

$$\Phi(x,y) = y^{n+1-m} \Delta^n \frac{u}{y}$$

with

$$u = \log \frac{x^2+(y-b)^2}{x^2+(y+b)^2} . \tag{82}$$

By a detailed investigation (cf. [15]) it follows in the case $n = m-1$, $m \in \mathbb{N}$:

$$\begin{cases} \Phi(x,0) = \varphi_1(x), \\ \Phi_x(x,0) = \varphi_2(x), \\ \Phi_y(x,0) = 0 \end{cases} \tag{83}$$

with

$$\varphi_1(x) = \frac{(-4)^m b[(m-1)!]^2}{(2m-1)[x^2+b^2]^{2m-1}} \sum_{\mu=0}^{m-1} (-1)^\mu \binom{2m-1}{2\mu} b^{2(m-1-\mu)} x^{2\mu} , \tag{84}$$

$$\varphi_2(x) = \frac{-(4)^m b[(m-1)!]^2}{[x^2+b^2]^{2m}} \sum_{\mu=0}^{m-1} (-1)^\mu \binom{2m}{2\mu+1} b^{2(m-1-\mu)} x^{2\mu+1} . \tag{85}$$

In the case $-m = n \in \mathbb{N}_o$ we get:

$$\Phi(x,0) = \Phi_x(x,0) = 0$$

$$\Phi_y(x,0) = \begin{cases} -\dfrac{4b}{x^2+b^2} & \text{for } m = 0, \\ 0 & \text{for } -m \in \mathbb{N}. \end{cases}$$

Using in (70) the generator

$$\gamma(x,t) = 2 \text{ arc tg } \frac{x-t}{b} - 2 \text{ arc tg } \frac{x+t}{b}$$

for the representation in the hyperbolic half-plane, we obtain for $n = m-1$, $m \in \mathbb{N}$:

$$X(x,0) = \varphi_1(x),$$

$$X_x(x,0) = \varphi_2(x),$$

$$X_y(x,0) = 0$$

with $\varphi_1(z)$ and $\varphi_2(z)$ according to (84) and (85) respectively. In the case $-m = n \in \mathbb{N}_o$ we have:

$$X(x,0) = X_x(x,0) = 0,$$

$$X_t(x,0) = \begin{cases} -\dfrac{4b}{x^2+b^2} & \text{for } m = 0, \\ 0 \text{ for } -m \in \mathbb{N} . \end{cases}$$

Summarizing we obtain the following

Theorem 13

a) Using in the elliptic and hyperbolic half-plane the generators

$$u = \log \frac{x^2+(y-b)^2}{x^2+(y+b)^2}, \quad b > 0,$$

and

$$\gamma = 2 \text{ arc tg } \frac{x-t}{b} - 2 \text{ arc tg } \frac{x+t}{b},$$

respectively, we obtain by

$$(86)\qquad \Psi(\xi,\eta) = \begin{cases} \left[y^{n+1-m}\triangle^{n}\frac{u}{y}\right]_{x=\xi,\ y=\frac{2}{3}\eta^{\frac{3}{2}}} & \text{for } \eta \geq 0 \\ \\ \left[t^{n+1-m}H^{n}\frac{\gamma}{t}\right]_{x=\xi,\ t=\frac{2}{3}(-\eta)^{\frac{3}{2}}} & \text{for } \eta \leq 0 \end{cases}$$

with

$$n = m-1 \quad \text{for} \quad m \in \mathbb{N},$$

$$n = -m \quad \text{for} \; -m \in \mathbb{N}_0$$

a fundamental solution in the large for the differential equation (54)

$$\eta\,\Psi_{\xi\xi} + \Psi_{\eta\eta} + \frac{6m-1}{2}\Psi_{\eta} = 0, \quad m \in \mathbb{Z}.$$

b) The solution (86) and the derivatives Ψ_ξ and Ψ_η are continuous in the whole plane up to the point

$$P_0\left(0,\left(\frac{3b}{2}\right)^{\frac{2}{3}}\right)$$

and it holds

$$\Psi(\xi,0) = \varphi_1(\xi),\ \Psi_\xi(\xi,0) = \varphi_2(\xi),\quad \Psi_\eta(\xi,0) = 0$$

for $n = m-1$, $m \in \mathbb{N}$, and

$$\Psi(\xi,0) = \Psi_\xi(\xi,0) = 0 \quad \text{for} \quad n = -m \in \mathbb{N}_0,$$

$$\Psi_\eta(\xi,0) = \begin{cases} \dfrac{-4b}{\xi^2+b^2} & \text{for} \quad m = 0, \\ \\ 0 \quad \text{for} & -m \in \mathbb{N}. \end{cases}$$

Here, the functions $\varphi_1(\xi)$ and $\varphi_2(\xi)$ are given by (84) and (85).

5) Generalized Stokes-Beltrami systems

By means of the results in Chapter I we may derive also representations of solutions for generalized Stokes-Beltrami systems. Let us illustrate this fact by the system

$$\eta^p w_z = w^*_z , \tag{87a}$$

$$\lambda\eta^p w_{\bar{z}} = -w^*_{\bar{z}} \tag{87b}$$

with

$$\eta = z-\bar{z} > 0, \quad p > 0, \quad \lambda \in \mathbb{C}, \quad \lambda(\lambda+1) \neq 0,$$

which is closely related to certain Stokes-Beltrami systems arising in connection with physical and technical problems. Thus, it is also possible to get real-valued solutions of those systems (cf. Theorem 16 and Theorem 17).

In this section by D we denote a simply connected domain of the upper half-plane. By a solution (w,w^*) of (87) we mean a pair of functions $w(z,\bar{z})$ and $w^*(z,\bar{z})$ which are twice continuously differentiable in D and satisfy the differential equations (87a) and (87b) in D.

If we differentiate (87a) and (87b) with respect to z and $\bar{z}$ respectively, by addition we get

$$\eta^p(w_z+w_{\bar{z}})_z = (w^*_z+w^*_{\bar{z}})_z$$

and

$$\lambda\eta^p(w_z+w_{\bar{z}})_{\bar{z}} = -(w^*_z+w^*_{\bar{z}})_{\bar{z}} .$$

Therefore, if (w,w^*) is a solution of (87), we obtain a further solution of this system by

$$W = w_z + w_{\bar{z}} , \quad W^* = w^*_z + w^*_{\bar{z}} . \tag{88}$$

A corresponding statement holds for

(89) $$W = \frac{\lambda w^*_z - w^*_{\bar{z}}}{\lambda \eta^p} , \quad W^* = \eta^p (w_z - \lambda w_{\bar{z}})$$

and

(90) $$\begin{cases} W = w_{zz} + w_{\bar{z}\bar{z}} + \frac{2p}{(\lambda+1)\eta} (w_z - \lambda w_{\bar{z}}) , \\ \\ W^* = w^*_{zz} + w^*_{\bar{z}\bar{z}} - \frac{2p}{(\lambda+1)\eta} (\lambda w^*_z - w^*_{\bar{z}}) \end{cases}$$

(cf. [20]).

Analogous to A. Weinstein's correspondence principle (cf. [110]) here we have the following assertion: If (w,w^*) is a solution of (87), we obtain a solution (W,W^*) of the system

(91a) $$\eta^{p+2} W_z = W^*_z ,$$

(91b) $$\tau \eta^{p+2} W_{\bar{z}} = -W^*_{\bar{z}}$$

with

$$\tau = \frac{\lambda+1+\lambda p}{\lambda+1+p} , \quad (\lambda+1+\lambda p)(\lambda+1+p) \neq 0$$

by

(92) $$W = \frac{w_z - \lambda w_{\bar{z}}}{\eta} ,$$

(93) $$W^* = \eta (w^*_z - \tau w^*_{\bar{z}}) - \frac{(\lambda+1)(p+1)(p+2)}{\lambda+1+p} w^* .$$

Similarly by

(94) $$W = \eta (w_z - \sigma^{-1} w_{\bar{z}}) + \frac{(\lambda+1)(p-1)(2-p)}{\lambda+1-\lambda p} w ,$$

(95) $$W^* = \frac{w^*_z - \lambda^{-1} w^*_{\bar{z}}}{\eta}$$

we get a solution fo the system

(96a) $$\eta^{p-2} w_z = w^*_z ,$$

(96b) $$\sigma \eta^{p-2} w_{\bar z} = -w^*_{\bar z}$$

with

$$\sigma = \frac{\lambda+1-\lambda p}{\lambda+1-p} , \quad (\lambda+1-\lambda p)(\lambda+1-p) \neq 0.$$

In these transitions from p to p+2 respectively p-2 the parameter λ is invariant only in the case $\lambda = 1$.
The functions w and w* of a solution of (87) satisfy the Euler equation [40]

(97) $$(\lambda+1)\eta w_{z\bar z} - p(w_z - \lambda w_{\bar z}) = 0,$$

(98) $$(\lambda+1)\eta w^*_{z\bar z} + p(\lambda w^*_z - w^*_{\bar z}) = 0,$$

which are transformed into the differential equations

(99) $$\eta^2 v_{z\bar z} + \frac{p(\lambda-1)}{\lambda+1}\eta v_{\bar z} + \frac{p(\lambda p-\lambda-1)}{(\lambda+1)^2} v = 0$$

and

(100) $$\eta^2 v^*_{z\bar z} + \frac{p(\lambda-1)}{\lambda+1}\eta v^*_{\bar z} + \frac{\lambda p(p+\lambda+1)}{(\lambda+1)^2} v^* = 0$$

by

(101) $$w = \eta^{\frac{-p}{\lambda+1}} v, \quad w^* = \eta^{\frac{\lambda p}{\lambda+1}} v^*,$$

respectively. The solutions of these differential equations defined in D can be obtained by means of Theorem I,6 if we set

$$\varphi(z) = z \quad \text{and} \quad \psi(z) = -z,$$

and if the parameters λ and p satisfy certain conditions. Namely, if

$$\frac{p(\lambda-1)}{\lambda+1} = n-m, \quad \frac{\lambda p(p+\lambda+1)}{(\lambda+1)^2} = n(m+1), \quad n,m \in \mathbb{N},$$

it follows (for $p > 0$)

$$\lambda = \frac{n}{m}, \quad p = n+m$$

and

$$\frac{p(\lambda p-\lambda-1)}{(\lambda+1)^2} = (n-1)m.$$

Applying the above-mentioned theorem, by suitable normalization we get the solutions

$$(102) \qquad v = \sum_{k=0}^{n-1} (-1)^k B_k^{n-1,m-1} \frac{g_1^{(k)}(z)}{\eta^{n-1-k}} + \sum_{k=0}^{m-1} B_k^{m-1,n-1} \frac{\overline{h_1^{(k)}(z)}}{\eta^{n-1-k}},$$

$$(103) \qquad v^* = \sum_{k=0}^{n} (-1)^k B_k^{n,m} \frac{g_2^{(k)}(z)}{\eta^{n-k}} + \sum_{k=0}^{m} B_k^{m,n} \frac{\overline{h_2^{(k)}(z)}}{\eta^{n-k}}$$

with

$$(104) \qquad B_k^{n,m} = \frac{(n+m-k)!}{k!(n-k)!}, \quad g_j(z),\ h_j(z) \in H(D), \quad j = 1,2.$$

Transforming by (101) and substituting into (87), it follows

$$ng_1(z) = -mg_2'(z) \quad \text{and} \quad h_1(z) = -h_2'(z).$$

Applying Theorem I,6 we obtain the solutions $v^* \equiv 0$ in the representation (103) by

$$g_2(z) = \sum_{\mu=0}^{n+m} a_\mu z^\mu, \quad h_2(z) = -\frac{m!}{n!} \sum_{\mu=0}^{n+m} \overline{a_\mu} z^\mu, \quad a_\mu \in \mathbb{C}.$$

If we consider the corresponding statement for (102) with $n-1$ and $m-1$ instead of n and m respectively, by inserting it follows $a_{n+m} = 0$.

Theorem 14

Let D be a simply connected domain of the upper half-plane.

a) For every solution (w,w^*) of the system (87)

$$\eta^p w_z = w^*_z \,, \quad \lambda\eta^p w_{\bar z} = -w^*_{\bar z}$$

with

$$\frac{p}{\lambda+1} = m \in \mathbb{N} \quad \text{and} \quad \frac{\lambda p}{\lambda+1} = n \in \mathbb{N}$$

defined in D there exist two functions $g(z)$ and $h(z) \in H(D)$, such that

$$(105) \qquad w = -\frac{m}{n}\sum_{k=0}^{n-1}(-1)^k B_k^{n-1,m-1}\frac{g^{(k+1)}(z)}{\eta^{n+m-1-k}} - \sum_{k=0}^{m-1} B_k^{m-1,n-1}\frac{\overline{h^{(k+1)}(z)}}{\eta^{n+m-1-k}} \,,$$

$$(106) \qquad w^* = \sum_{k=0}^{n}(-1)^k B_k^{n,m}\eta^k g^{(k)}(z) + \sum_{k=0}^{m} B_k^{m,n}\eta^k\, \overline{h^{(k)}(z)} \,.$$

b) Conversely, for each pair of functions $g(z)$, $h(z) \in H(D)$ (105) and (106) represent a solution of (87) in D.

c) For every given solution (w,w^*) the generators $g(z)$ and $h(z)$ are not uniquely determined. We get the most general pair of generators $\tilde g(z)$ and $\tilde h(z)$ by

$$(107) \qquad \tilde g = g + \sum_{\mu=0}^{n+m-1} a_\mu z^\mu \,, \quad \tilde h = h - \frac{m!}{n!}\sum_{\mu=0}^{n+m-1}\overline{a_\mu} z^\mu \,, \quad a_\mu \in \mathbb{C}.$$

For the elliptic differential equation

$$(108) \qquad \eta^2 v_{z\bar z} + A\eta v_{\bar z} + Bv = 0, \quad A,B \in \mathbb{C},$$

a general representation theorem for the solutions is not known. Thus, it is not possible to derive a corresponding assertion for arbitrary

values $p > 0$, $\lambda \in \mathbb{C}$. However, as in the case of the differential equation (I,175) we may determine certain classes of solutions in which homogeneous polynomials in z and $\bar{z}$ of arbitrary degree arise.
Using the particular solutions

$$v = \eta^{\sigma} \ , \quad \sigma^2 + \sigma(A-1) = B,$$

of (108), we set

$$v = \eta^{\sigma} z^m f(\xi), \quad \xi = \frac{\bar{z}}{z} \ , \quad m \in \mathbb{N}_0,$$

then, $f(\xi)$ has to satisfy the hypergeometric differential equation

$$\xi(\xi-1)f'' + [(\alpha+\beta+1)\xi - \gamma]f' + \alpha\beta f = 0 \tag{109}$$

with

$$\alpha = \sigma \ , \quad \beta = -m, \quad \gamma = 1-A-\sigma-m. \tag{110}$$

With respect to (99) we set

$$A = \frac{p(\lambda-1)}{\lambda+1} \ , \quad B = \frac{p(\lambda p-\lambda-1)}{(\lambda+1)^2} \ .$$

Then, it follows

$$\sigma_1 = \frac{p}{\lambda+1} \ , \quad \sigma_2 = 1 - \frac{\lambda p}{\lambda+1}$$

and

$$v_1 = \eta^{\frac{p}{\lambda+1}} z^m F\left(\frac{p}{\lambda+1} \ , -m, \ 1-m-\frac{\lambda p}{\lambda+1} \ ; \frac{\bar{z}}{z}\right),$$

$$v_2 = \eta^{\frac{\lambda+1-\lambda p}{\lambda+1}} z^m F\left(1-\frac{\lambda p}{\lambda+1} \ , -m, \ \frac{p}{\lambda+1} - m; \frac{\bar{z}}{z}\right),$$

$$m \in \mathbb{N}_0, \quad \left(1-m-\frac{\lambda p}{\lambda+1}\right)_m \neq 0 \ \text{ resp. } \left(\frac{p}{\lambda+1} - m\right)_m \neq 0,$$

if $F(\alpha,\beta,\gamma;\xi)$ denotes the hypergeometric function (cf. e.g. [87]).

In view of (100) we set

$$A = \frac{p(\lambda-1)}{\lambda+1}, \quad B = \frac{\lambda p(p+\lambda+1)}{(\lambda+1)^2},$$

it follows

$$\sigma_1 = 1 + \frac{p}{\lambda+1}, \quad \sigma_2 = -\frac{\lambda p}{\lambda+1},$$

and we obtain the solutions

$$v_1^* = \eta^{\frac{\lambda+1+p}{\lambda+1}} z^m F\left(1 + \frac{p}{\lambda+1}, -m, -m-\frac{\lambda p}{\lambda+1}; \frac{\bar{z}}{z}\right),$$

$$v_2^* = \eta^{\frac{-\lambda p}{\lambda+1}} z^m F\left(\frac{-\lambda p}{\lambda+1}, -m, 1 + \frac{p}{\lambda+1} - m; \frac{\bar{z}}{z}\right),$$

$$m \in \mathbb{N}_o, \quad \left(-m - \frac{\lambda p}{\lambda+1}\right)_m \neq 0 \quad \text{resp.} \quad \left(1 + \frac{p}{\lambda+1} - m\right)_m \neq 0.$$

With respect to the powers of z and $\bar{z}$ we set for a solution of (87)

(111a) $$w = c_1 z^{n+1} F\left(\frac{p}{\lambda+1}, -n-1, -n-\frac{\lambda p}{\lambda+1}; \frac{\bar{z}}{z}\right),$$

(111b) $$w^* = z^n \eta^{p+1} F\left(1 + \frac{p}{\lambda+1}, -n, -n-\frac{\lambda p}{\lambda+1}; \frac{\bar{z}}{z}\right)$$

respectively

(112a) $$w = z^m \eta^{1-p} F\left(1 - \frac{\lambda p}{\lambda+1}, -m, \frac{p}{\lambda+1} - m; \frac{\bar{z}}{z}\right),$$

(112b) $$w^* = c_2 z^{m+1} F\left(-\frac{\lambda p}{\lambda+1}, -m-1, \frac{p}{\lambda+1} - m; \frac{\bar{z}}{z}\right)$$

with arbitrary constants c_1 and c_2. Substituting into (87), we obtain solutions if

$$c_1 = \frac{n+1+p}{n+1}, \quad c_2 = \frac{m+1-p}{m+1}.$$

Theorem 15

If $m,n \in \mathbb{N}$, $p > 0$, $\lambda, C_1, C_2 \in \mathbb{C}$ with $\lambda(\lambda+1) \neq 0$ and

$$\left(-n - \frac{\lambda p}{\lambda+1}\right)_{n+1} \neq 0, \quad \left(\frac{p}{\lambda+1} - m\right)_m \neq 0,$$

then

$$(113) \qquad w = C_1(n+1+p)z^{n+1} F\left(\frac{p}{\lambda+1}, -n-1, -n - \frac{\lambda p}{\lambda+1}; \frac{\bar{z}}{z}\right) + $$

$$+ C_2(m+1)z^m \eta^{1-p} F\left(1 - \frac{\lambda p}{\lambda+1}, -m, \frac{p}{\lambda+1} - m; \frac{\bar{z}}{z}\right),$$

$$(114) \qquad w^* = C_1(n+1)z^n \eta^{p+1} F\left(1 + \frac{p}{\lambda+1}, -n, -n - \frac{\lambda p}{\lambda+1}; \frac{\bar{z}}{z}\right) + $$

$$+ C_2(m+1-p)z^{m+1} F\left(\frac{-\lambda p}{\lambda+1}, -m-1, \frac{p}{\lambda+1} - m; \frac{\bar{z}}{z}\right)$$

represents a solution of (87).

In connection with physical and technical problems frequently one is led to Stokes-Beltrami systems. For instance, A. Weinstein treated in [110] the system

$$(115a) \qquad y^p \Phi_x = \Psi_y ,$$

$$(115b) \qquad y^p \Phi_y = -\Psi_x$$

for $p > 0$ and developed the generalized axially symmetric potential theory. Among other things A. Weinstein considered applications in hydrodynamics and electrostatics, in connection with the torsion of shafts in revolution and transonic flow. Moreover, matrix transformations for Stokes-Beltrami equations have been the subject of several papers in recent years (cf. e.g. [91,96-98]). In this context invariance and reduction properties were investigated and applications in gasdynamics and aligned magneto-gasdynamics were treated. Apart from (115) (cf. e.g. [91,96,97]), for instance, C. Rogers and J.G. Kingston [98] dealt with the system

(116a) $$\Phi_r = -r^{-1}(r \sin \varphi)^{-p}\, \Psi_\varphi \,,$$

(116b) $$\Psi_r = r^{-1}(r \sin \varphi)^{p}\, \Phi_\varphi, \quad p \in \mathbb{R}.$$

By means of the results for the system (87) we may derive the corresponding real-valued solutions for the systems (115) and (116) without difficulties. Since the system (116) can be reduced to (115), in the following we consider only (115). Transforming this system by

(117) $$\Phi(x,y) = (2i)^p w(z,\bar{z}), \quad \Psi(x,y) = -iw^*(z,\bar{z}), \quad z = x+iy,$$

we obtain (87) with $\lambda = 1$. First, we have

$$p = 2n, \quad n \in \mathbb{N},$$

and for the real-valued solutions we get with $2f = h-g$ by $2\Phi = \Phi+\bar{\Phi}$ and $2\Psi = \Psi+\bar{\Psi}$

(118) $$\Phi(x,y)=(-4)^n\left[\sum_{k=0}^{n-1} \frac{B_k^{n-1}}{\eta^{2n-1-k}}\left[(-1)^k f^{(k+1)}(z)-\overline{f^{(k+1)}(z)}\,\right]\right]_{z=x+iy},$$

(119) $$\Psi(x,y) = i\left[\sum_{k=0}^{n} B_k^n \eta^k \left[(-1)^k f^{(k)}(z)-\overline{f^{(k)}(z)}\right]\right]_{z=x+iy}$$

with

(120) $$B_k^n = \frac{(2n-k)!}{k!(n-k)!}\,.$$

Setting

$$f(x) = X(x,y) + iY(x,y), \quad X,Y \text{ real-valued},$$

and imploying again the operators

(121) $$D_x = \frac{\partial}{\partial x}\,, \quad D_y = \frac{\partial}{\partial y}\,,$$

it follows by the Cauchy-Riemann equations

$$X_x = Y_y \ , \quad X_y = -Y_x$$

and suitable normalization

(122) $$\Phi(x,y) = 2\sum_{k=0}^{n-1} C_k^{n-1} \frac{D_x D_y^k Y}{y^{2n-1-k}} \ ,$$

(123) $$\Psi(x,y) = \sum_{k=0}^{n} C_k^n y^k D_y^k Y$$

with

(124) $$C_k^n = \frac{(2n-k)!(-2)^k}{k!(n-k)!} \ .$$

Theorem 16

Let D be a simply connected domain of the upper x,y-half-plane.

a) For every solution (Φ,Ψ) of the system

(125a) $$y^{2n} \Phi_x = \Psi_y \ ,$$

(125b) $$y^{2n} \Phi_y = -\Psi_x \ , \quad n \in \mathbb{N},$$

defined in D, there exist a harmonic function Y(x,y) in D, such that

(126) $$\Phi(x,y) = 2\sum_{k=0}^{n-1} C_k^{n-1} \frac{D_x D_y^k Y}{y^{2n-1-k}} \ ,$$

(127) $$\Psi(x,y) = \sum_{k=0}^{n} C_k^n y^k D_y^k Y$$

with C_k^n, D_x, D_y according to (124) and (121) respectively.

b) Conversely, for each harmonic function Y(x,y) in D (126) and (127) represent a solution of the system (125) in D.

Applying the results of Theorem 15 with $\lambda = 1$ and $C_1 = 0$, now we consider the functions

$$\Phi_1 = y^{1-p}\left[z^m F\left(1 - \frac{p}{2}, -m, \frac{p}{2} - m; \frac{\bar{z}}{z}\right)\right]_{z=x+iy},$$

$$\Psi_1 = \left[z^{m+1} F\left(-\frac{p}{2}, -m-1, \frac{p}{2} - m; \frac{\bar{z}}{z}\right)\right]_{z=x+iy}.$$

If m is an odd positive integer, it follows with $m = 2n-1$ by

$$(128) \qquad (1+\xi)^{-\alpha} F\left(\frac{\alpha}{2}, \frac{\alpha+1}{2}, 1+\alpha-\beta; \frac{4\xi}{(1+\xi)^2}\right) = F(\alpha,\beta,1+\alpha-\beta;\xi)$$

(cf. e.g. [87])

$$\Phi_1 = C_1 y^{1-p} x^{2n-1} F\left(\frac{1}{2} - n, 1-n, \frac{3-p}{2}; -\frac{y^2}{x^2}\right),$$

$$\Psi_1 = C_2 x^{2n} F\left(\frac{1}{2} - n, -n, \frac{1-p}{2}; -\frac{y^2}{x^2}\right),$$

where C_1 and C_2 denote certain constants. By inserting

$$\Phi = C\Phi_1, \quad \Psi = \Psi_1, \quad C \text{ constant},$$

into (115), we obtain by suitable normalization

$$\Phi = 2n\, y^{1-p} x^{2n-1} F\left(\frac{1}{2} - n, 1-n, \frac{3-p}{2}; -\frac{y^2}{x^2}\right),$$

$$\Psi = (p-1) x^{2n} F\left(\frac{1}{2} - n, -n, \frac{1-p}{2}; -\frac{y^2}{x^2}\right),$$

$$n \in \mathbb{N}, \quad \left(\frac{1-p}{2}\right)_n \neq 0.$$

By a corresponding procedure we get solutions in the case $m = 2n$ and for $\lambda = 1$ and $C_2 = 0$.

Theorem 17

If $n,s \in \mathbb{N}$, $m,t \in \mathbb{N}_o$, $p > 0$ and

$$\left(\frac{1-p}{2}\right)_n \neq 0, \quad \left(\frac{1-p}{2}\right)_m \neq 0,$$

then, the following four pairs of functions (Φ,Ψ) represent solutions of the system (115)

$$y^p\Phi_x = \Psi_y , \quad y^p\Phi_y = -\Psi_x :$$

1) $$\Phi = 2n\, y^{1-p}\, x^{2n-1}\, F\left(\frac{1}{2} - n,\ 1-n,\ \frac{3-p}{2}\ ;\ -\frac{y^2}{x^2}\right),$$

$$\Psi = (p-1)x^{2n}\, F\left(\frac{1}{2} - n,\ -n,\ \frac{1-p}{2}\ ;\ -\frac{y^2}{x^2}\right).$$

2) $$\Phi = (2m+1)y^{1-p}\, x^{2m}\, F\left(\frac{1}{2} - m,\ -m,\ \frac{3-p}{2}\ ;\ -\frac{y^2}{x^2}\right),$$

$$\Psi = (p-1)x^{2m+1}\, F\left(-\frac{1}{2} - m,\ -m,\ \frac{1-p}{2}\ ;\ -\frac{y^2}{x^2}\right).$$

3) $$\Phi = (p+1)x^{2t+1}\, F\left(-\frac{1}{2} - t,\ -t,\ \frac{1+p}{2}\ ;\ -\frac{y^2}{x^2}\right),$$

$$\Psi = (2t+1)y^{p+1}\, x^{2t}\, F\left(\frac{1}{2} - t,\ -t,\ \frac{3+p}{2}\ ;\ -\frac{y^2}{x^2}\right).$$

4) $$\Phi = (p+1)x^{2s}\, F\left(\frac{1}{2} - s,\ -s,\ \frac{1+p}{2}\ ;\ -\frac{y^2}{x^2}\right),$$

$$\Psi = 2s\, y^{p+1}\, x^{2s-1}\, F\left(\frac{1}{2} - s,\ 1-s,\ \frac{3+p}{2}\ ;\ -\frac{y^2}{x^2}\right).$$

For the solutions of the Euler equation

$$\eta\, w_{z\bar{z}} + \nu\, w_z - \mu\, w_{\bar{z}} = 0, \quad \eta = z-\bar{z}, \quad \nu,\mu \in \mathbb{C}, \tag{129}$$

there is a number of functional-differential-relations. Some of these statements which may easily be verified are summarized in the following theorem. Here, we denote the set of solutions of (129) defined in D by $F_{\mu,\nu}(D)$.

Theorem 18

If $w \in F_{\mu,\nu}(D)$, then:

$$w_z + w_{\bar{z}} \in F_{\mu,\nu}(D), \tag{130}$$

$$w_z \in F_{\mu-1,\nu}(D), \tag{131}$$

$$w_{\bar{z}} \in F_{\mu,\nu-1}(D), \tag{132}$$

$$\eta w_{\bar{z}} + (\mu+\nu+1)w \in F_{\mu+1,\nu}(D), \tag{133}$$

$$\eta w_z - (\mu+\nu+1)w \in F_{\mu,\nu+1}(D), \tag{134}$$

$$\eta w_z - \mu w \in F_{\mu-1,\nu+1}(D), \tag{135}$$

$$\eta w_{\bar{z}} + \nu w \in F_{\mu+1,\nu-1}(D), \tag{136}$$

$$\eta^{-(\mu+\nu)}(\nu w_z - \mu w_{\bar{z}}) \in F_{-\nu,\, -\mu}(D), \tag{137}$$

$$\eta^{-(\nu+\mu+1)} w \in F_{-(\nu+1),-(\mu+1)}(D), \tag{138}$$

$$\eta^{-1}(\nu w_z - \mu w_{\bar{z}}) \in F_{\mu-1,\nu-1}(D), \tag{139}$$

$$\eta\,[(\mu+1)w_z - (\nu+1)w_{\bar{z}}] - (\nu+\mu+1)(\nu+\mu+2)w \in F_{\mu+1,\nu+1}(D). \tag{140}$$

Applying these results we are led to a number of systems of first-order differential equations whose solutions may be represented by differential operators if the parameters μ and ν are integers. Let us illustrate this fact by some examples. Proceeding from the representation of the solutions of (I,56) summarized in Theorem I,6 we set

$$\varphi(z) = z \quad \text{and} \quad \psi(z) = -z,$$

and by suitable transformation we obtain the solutions of the Euler equation in a simply connected domain D by

$$(141) \qquad w = X(\nu,\mu;g,\overline{f}), \quad \mu,\nu \in \mathbb{N}_o,$$

and

$$(142) \qquad w = \eta^{1+\lambda+\mu} X(-\mu-1, -\nu-1;g,\overline{f}), \quad -\mu,-\nu \in \mathbb{N}.$$

g(z) and h(z) are arbitrary holomorphic functions in D, whereas the operator X is defined by

$$(143) \quad X(\nu,\mu;g,\overline{f}) = \sum_{k=0}^{\nu} \frac{(-1)^k(\nu+\mu-k)!}{k!(\nu-k)!}\eta^k g^{(k)}(z) + \sum_{k=0}^{\mu} \frac{(\nu+\mu-k)!}{k!(\mu-k)!}\eta^k \overline{f^{(k)}(z)}.$$

For example, if we proceed from the relations (137) and (130), we obtain the above-considered Stokes-Beltrami system (87) with

$$p = \mu + \nu, \quad \lambda = \frac{\nu}{\mu}, \quad \mu,\nu \in \mathbb{N}.$$

Proceeding from $w \in F_{\mu,\nu}(D)$, by (135) we have

$$\eta w_z - \mu w \in F_{\mu-1,\nu+1}(D).$$

Using $w^* \in F_{\mu-1,\nu+1}(D)$, by (136) it follows

$$\eta w^*_{\overline{z}} + (\nu+1)w^* \in F_{\mu,\nu}(D).$$

Thus, we are led to the system

$$(144) \qquad \begin{cases} \eta w_z - \mu w = w^* \\ \eta w^*_{\overline{z}} + (\nu+1)w^* = -\mu(\nu+1)w. \end{cases}$$

If $\mu \in \mathbb{N}$, $\nu \in \mathbb{N}_o$, by (141) we obtain the solution

$$\begin{cases} w = X(\nu,\mu;-g,\bar{f}), \\ w^* = X(\nu+1,\mu-1;\mu(\nu+1)g,-\bar{f}). \end{cases}$$

If μ and ν are negative integers with $\mu \leq -1$, $\nu \leq -2$, by (142) we get the solution

$$\begin{cases} w = \eta^{\mu+\nu+1}X(-\mu-1,-\nu-1;-g,\bar{f}), \\ w^* = \eta^{\mu+\nu+1}X(-\mu,-\nu-2;\mu(\nu+1)g,-\bar{f}). \end{cases}$$

Using $w \in F_{\mu,\nu}(D)$ and $w^* \in F_{\mu-1,\nu}(D)$, by (131), (130), and (133) one is led to the system

$$\begin{cases} w_z = w^*_z + w^*_{\bar{z}} \\ \eta w^*_{\bar{z}} + (\mu+\nu)w^* = \mu w. \end{cases}$$

In this case we obtain the solutions defined in D by

$$\begin{cases} w = X(\nu,\mu;g,\bar{f}), \\ w^* = X(\nu,\mu-1;\mu g,\bar{f}), \end{cases}$$

for $\mu \in \mathbb{N}$, $\nu \in \mathbb{N}_0$ and by

$$\begin{cases} w = \eta^{\mu+\nu+1}X(-\mu-1,-\nu-1;g',-\bar{f}), \\ w^* = \eta^{\mu+\nu}X(-\mu,-\nu-1;\mu g,\bar{f}) \end{cases}$$

for $-\mu,-\nu \in \mathbb{N}$.

6) The iterated equation of generalized axially symmetric potential theory

The iterated equation of generalized axially symmetric potential theory (cf. [110]) runs

$$\hat{X}_k^m f = 0, \quad k \in \mathbb{R}, \quad m \in \mathbb{N}, \tag{145}$$

with

$$\hat{X}_k = \frac{\partial^2}{\partial x^2} + \frac{\partial^2}{\partial y^2} + \frac{k}{y}\frac{\partial}{\partial y} .$$

This differential equation arises in a number of physical and technical problems and was treated, for instance, by A. Weinstein [110,112], L. E. Payne [90], R.J. Weinacht [109], and P.C. Burns [38]. Setting especially $k = 0$, we get the differential equation of the two-dimensional polyharmonic functions.
By $z = x+iy$, $f(x,y) = w(z,\bar{z})$, it follows

$$X_k^m w = 0 \tag{146}$$

with

$$\frac{1}{4}\hat{X}_k = X_k = \frac{\partial^2}{\partial z \partial \bar{z}} + \frac{k}{2\eta}\left(\frac{\partial}{\partial \bar{z}} - \frac{\partial}{\partial z}\right), \quad \eta = z-\bar{z},$$

and by $w = \eta^{-\frac{k}{2}} v$ we obtain the differential equation

$$Y^m_{\frac{k}{2}-1} v = 0$$

with

$$Y_{\frac{k}{2}-1} = \frac{\partial^2}{\partial z \partial \bar{z}} + \frac{1}{\eta^2}\left(\frac{k}{2} - 1\right)\frac{k}{2} .$$

For $m = 1$ and $\frac{k}{2} \in \mathbb{Z}$ we get a particular case of the differential equation (I,67); thus, the possibility offers to determine also the solutions of the corresponding iterated equation. In this connection we may confine ourselves to the differential equation

(147) $$Y_n^m v = 0, \quad n \in \mathbb{N}_0 .$$

Let D be a simply connected domain of the upper half-plane. First, for $m \geq 2$ we set

$$V_1 = Y_n^{m-1} v,$$

and obtain the differential equation

$$Y_n V_1 = 0$$

whose solutions defined in D are given by

$$V_1 = H_n g + \overline{H_n h}, \quad g(z), h(z) \in H(D),$$

with

(148) $$H_n = \sum_{k=0}^{n} \frac{A_k^n}{\eta^{n-k}} \frac{\partial^k}{\partial z^k}$$

(cf. Theorem I,9). Then, by

$$V_2 = Y_n^{m-2} v,$$

it follows

$$Y_n V_2 = V_1 .$$

In order to determine a particular solution of this inhomogeneous differential equation we set

$$V_2 = \tau u, \quad \tau = z + \bar{z} .$$

By

$$d = \frac{\partial}{\partial z} + \frac{\partial}{\partial \bar{z}}$$

it follows that

$$Y_n(\tau u) = du,$$

if u is a solution of

$$Y_n u = 0 . \tag{149}$$

By $Y_n d = dY_n$ we get

$$Y_n(du) = d(Y_n u) = 0.$$

Therefore, if

$$u = H_n g + \overline{H_n h}$$

is a solution of (149) in D, also

$$du = H_n g' + \overline{H_n h'}$$

represents a solution of this differential equation in D. Thus, the solutions of $Y_n^2 v = 0$ in D are given by

$$v = H_n g_o + \overline{H_n h_o} + \tau [H_n g_1 + \overline{H_n h_1}],$$

$$g_o, g_1, h_o, h_1 \in H(D).$$

If we contunue in this way we obtain the solutions of (147) in D by

$$v = \sum_{\sigma=0}^{m-1} \tau^{\sigma} v_{\sigma}$$

with

$$v_{\sigma} = H_n g_{\sigma} + \overline{H_n h_{\sigma}} , \quad g_{\sigma}(z), h_{\sigma}(z) \in H(D),$$

(cf. also [90]).
For a given solution v of (147) the function

$$d^{m-1} v_{m-1} = \frac{1}{(m-1)!} Y_n^{m-1} v \tag{150}$$

is uniquely determined. In general, for a given solution v the functions $d^{2(m-1)-\sigma} v_{\sigma}$ are uniquely determined and can be obtained recursively by the relations

$$(151)\qquad (m-\mu)!\,d^{m+\mu-2}v_{m-\mu} = d^{2(\mu-1)}Y_n^{m-\mu}v - 4d^{2(\mu-2)}Y_n^{m-\mu+1}v -$$

$$- d^{2(\mu-2)}Y_n^{m-\mu}\Big(\sum_{\sigma=m-\mu+1}^{m-1} \tau^{\sigma} d^2 v_{\sigma}\Big),$$

$$\mu = 2,3,\ \ldots,\ m-1,$$

as can be shown by induction. Applying Theorem I,9, we get the following general representation theorem (cf. [21]) for the solutions of (147).

Theorem 19

a) For every solution of the differential equation (147)

$$Y_n^m v = 0, \quad m \geq 2,$$

defined in D, there exist 2m functions $g_{\sigma}(z), h_{\sigma}(z) \in H(D)$, such that

$$(152)\qquad v = \sum_{\sigma=0}^{m-1} \tau^{\sigma}[H_n g_{\sigma} + \overline{H_n h_{\sigma}}].$$

b) Conversely, for arbitrary functions $g_{\sigma}(z), h_{\sigma}(z) \in H(D)$, $\sigma = 0,1,\ \ldots,\ m-1$, (152) represents a solution of (147) in D.

c) For every given solution v of (147) in D the functions $g_{\sigma}^{(2(m+n)-1-\sigma)}(z)$, $h_{\sigma}^{(2(m+n)-1-\sigma)}(z)$ are uniquely determined by

$$(153)\qquad g_{\sigma}^{(2(m+n)-1-\sigma)}(z) = \frac{P^{n+1}d^{2(m-1)-\sigma}v_{\sigma}}{\eta^{2n+2}},$$

$$(154)\qquad h_{\sigma}^{(2(m+n)-1-\sigma)}(z) = \frac{P^{n+1}\overline{d^{2(m-1)-\sigma}v_{\sigma}}}{\eta^{2n+2}}$$

with $P = \eta^2 \frac{\partial}{\partial z}$ and $d^{2(m-1)-\sigma}v_{\sigma}$ according to (150) and (151) respectively. In this case the generators $g_{\sigma}(z)$ and $h_{\sigma}(z)$ are not uniquely determined. We obtain the most general generators $\tilde{g}_\sigma(z)$ and $\tilde{h}_\sigma(z)$ by

$$\tilde{g}_\sigma(z) = g_\sigma(z) + p_\sigma(z), \quad \tilde{h}_\sigma(z) = h_\sigma(z) + q_\sigma(z),$$

where $p_\sigma(z)$ and $q_\sigma(z)$ denote polynomials in z of degree $2(m+n-1)-\sigma$ which satisfy the condition

$$(155) \qquad \sum_{\sigma=0}^{m-1} \tau^\sigma [H_n p_\sigma + \overline{H_n g_\sigma}] \equiv 0.$$

For instance, this condition is satisfied by

$$p_\sigma(z) = \sum_{\mu=0}^{2n} a_{\sigma\mu} z^\mu, \quad q_\sigma(z) = (-1)^{n+1} \sum_{\mu=0}^{2n} \overline{a_{\sigma\mu}} z^\mu, \quad a_{\sigma\mu} \in \mathbb{C}.$$

Those solutions (147) which can be represented only by $g_\sigma(z)$ have a special property which shall be pointed out. In the case $m = 1$ the generator $g_o(z)$ is uniquely determined by

$$(156) \qquad g_o(z) = \frac{(-1)^n \overline{p^{n-}_v}}{(2n)!}$$

(cf. (I,74)). A corresponding statement is not valid for $m \geq 2$. For example, in the case $m = 2$, $n = 1$ by

$$(157) \quad v = H_1 g_o + \tau H_1 g_1 \quad \text{with} \quad g_o(z) = -2Cz, \quad g_1(z) = C, \quad C \neq 0,$$

we obtain a solution which is identically zero.

The real-valued solutions of (147) are important for applications. Considering (152), we get these solutions by

$$(158) \qquad v = \sum_{\sigma=0}^{m-1} \tau^\sigma [H_n f_\sigma + \overline{H_n f_\sigma}], \quad f_\sigma(z) \in H(D).$$

Proceeding from (158), we may derive another simple representation for the real-valued solutions in which harmonic functions arise as generators and only derivatives with respect to y appear. Let D^* be a simply connected domain of the upper x,y-half-plane. We set

$$d_y = \frac{1}{y}\frac{\partial}{\partial y}$$

and suppose that $u_\sigma(x,y)$, $\sigma = 0,1, \ldots, m-1$, are arbitrary harmonic functions in D*. Then we obtain all real-valued solutions of (147) in D* by

$$v = y^{n+1} \sum_{\sigma=0}^{m-1} x^\sigma d_y^n \left(\frac{u_\sigma}{y}\right) ;$$

the solutions of $\hat{X}^m_{-2n} w = 0$ are given by

$$w = y^{2n+1} \sum_{\sigma=0}^{m-1} x^\sigma d_y^n \left(\frac{u_\sigma}{y}\right), \quad n \in \mathbb{N}_0 , \tag{159}$$

and we get the solutions of $\hat{X}^m_{2(n-1)} w = 0$ by

$$w = y \sum_{\sigma=0}^{m-1} x^\sigma d_y^{n-1} \left(\frac{u_\sigma}{y}\right), \quad n \in \mathbb{N} . \tag{160}$$

These results reduce with n = 0 in (159) and n = 1 in (160) to the known representation

$$w = \sum_{\sigma=0}^{m-1} x^\sigma u_\sigma$$

of the two-dimensional polyharmonic functions.

Applying the relations (153) and (154) we may derive general expansion theorems (cf. Chapter I,2b) for the solutions of (147) in the neighbourhood of isolated singularities. Here, we obtain the following result (cf. [21], Theorem 3).

<u>Theorem 20</u>

Let v be a complex-valued solution of (147) in $\dot{U}(z_0)$ with an isolated sinfularity at z_0. Then, v can be represented in $\dot{U}(z_0)$ by

$$v = \sum_{\sigma=0}^{m-1} \tau^\sigma [H_n g_\sigma + \overline{H_n h_\sigma}] \tag{161}$$

with the generators

$$g_\sigma(z) = g_\sigma^*(z) + S_\sigma(z)\log(z-z_o),$$

$$h_\sigma(z) = h_\sigma^*(z) + T_\sigma(z)\log(z-z_o),$$

where $g_\sigma^*(z)$ and $h_\sigma^*(z)$ are holomorphic and unique functions in $\dot{U}(z_o)$, whereas $S_\sigma(z)$ and $T_\sigma(z)$ represent polynomials in z of degree $2(m+n-1)-\sigma$ which satisfy the condition

$$(162)\qquad \sum_{\sigma=0}^{m-1} \tau^\sigma[H_n S_\sigma - \overline{H_n T_\sigma}] \equiv 0 .$$

For instance, this condition is satisfied by

$$S_\sigma(z) = \sum_{\mu=0}^{2n} a_{\sigma\mu} z^\mu , \quad T_\sigma(z) = (-1)^n \sum_{\mu=0}^{2n} \overline{a_{\sigma\mu}} z^\mu , \quad a_{\sigma\mu} \in \mathbb{C}.$$

If such a solution may be represented only by $g_\sigma(z)$, $\sigma = 0,1, \ldots, m-1$, by $T_\sigma(z) \equiv 0$ the condition (162) reduces to

$$(163)\qquad \sum_{\sigma=0}^{m-1} \tau^\sigma H_n S_\sigma \equiv 0.$$

In the case $m = 1$ it follows $S_o(z) \equiv 0$ by (156), whereas the polynomials $S_\sigma(z)$ are not identically zero for $m \geq 2$, as can be seen from (157). Thus, in the case $m = 1$ the generator $g_o(z)$ reduces to a Laurent series (cf. Theorem I,8), whereas, apart from the Laurent series $g_\sigma^*(z)$, logarithmic terms may arise for $m \geq 2$. By means of (161) it follows that the solutions v generally have the representation

$$v = \sum_{\sigma=0}^{m-1} \tau^\sigma\{[H_n S_\sigma]\log(z-z_o) + \overline{[H_n T_\sigma]\log(z-z_o)}\} + \sum ,$$

where $\sum$ represents a function free of logarithmic terms. Therefore, by $T_\sigma(z) \equiv 0$ and (163) it follows that the solutions v are free of logarithmic terms if they may be represented only by the generators $g_\sigma(z)$.

Applying the results of Chapter I corresponding assertions may be derived for the solutions of more general iterated differential equations of the type

$$Z^m v = 0, \quad m \in \mathbb{N}, \quad m \geq 2,$$

if Z denotes an operator of the form

$$Z = \frac{\partial^2}{\partial z \partial \bar{z}} + A(\eta)\frac{\partial}{\partial \bar{z}} + B(\eta), \quad \eta = z-\bar{z} .$$

For instance, this is the case for

$$Z_1 = \frac{\partial^2}{\partial z \partial \bar{z}} + \left[\frac{n(n+1)}{\eta^2} + \Phi'(\eta)\right] ,$$

with $\Phi(\eta)$ according to (I,144). Using Theorem I,18 we obtain the solutions of $Z_1 v = 0$ in D by

$$v = \frac{\eta^n}{C_1 + C_2\eta^{2n+1}} \left\{ C_2\eta^{n+1}[H_{n+1}g + \overline{H_{n+1}h}] + C_1\eta^{-n}[H_{n-1}g'' + \overline{H_{n-1}h''}]\right\},$$

$$g(z), h(z) \in H(D),$$

where H_n is given by (148). By

$$Z_2 v = 0$$

with

$$Z_2 = \frac{\partial^2}{\partial z \partial \bar{z}} + \frac{n-n^*}{\eta}\frac{\partial}{\partial \bar{z}} + \frac{n(n^*+1)}{\eta^2} , \qquad n, n^* \in \mathbb{N}_0 ,$$

we get a particular case of the differential equation (I,56), and a representation of the solutions can be found by means of Theorem I,6. Also here we have $dZ_\nu = Z_\nu d$, $\nu = 1,2$, and we obtain the solutions of the iterated differential equations $Z_\nu^m v = 0$ by

$$v = \sum_{\sigma=0}^{m-1} \tau^\sigma v_\sigma$$

with $Z_\nu v_\sigma = 0$, $\nu = 1,2$.

REFERENCES

[1] Ames, N.F.
Non Linear Partial Differential Equations in Engineering. Vol.II, Academic Press, London-New York: 1972.

[2] Barnard, T.W.
2Np Ultrashort Light Pulses. Phys. Rev., A7, 1, 373-376 (1973).

[3] Bauer, K.W.
Über die Lösungen der elliptischen Differentialgleichung $(1\pm z\bar{z})^2 w_{z\bar{z}} + \lambda w = 0$. Journ. Reine u. Angew. Math., 221, Teil I: S.48-84; Teil II: S.176-196 (1966).

[4] ---
Über eine der Differentialgleichung $(1\pm z\bar{z})^2 w_{z\bar{z}} \pm n(n+1)w=0$ zugeordnete Funktionentheorie. Bonner Math. Schriften, Nr.23, 1-98 (1965).

[5] ---
Über eine Differentialgleichung zweiter Ordnung mit zwei unabhängigen komplexen Variablen. Monatsh. f. Math., 70, 385-418 (1966).

[6] ---
Über eine Klasse verallgemeinerter Cauchy-Riemann'scher Differentialgleichungen. Math. Z., 100, 17-28 (1967).

[7] ---
Über eine Klasse homogener partieller Differentialgleichungen gerader Ordnung. Arch. d. Math., XVIII, 430-437 (1967).

[8] ---
Über die Lösung der inhomogenen elliptischen Differentialgleichung $(1+\varepsilon z\bar{z})^2 w_{z\bar{z}} + \varepsilon n(n+1)w = \Phi(z,\bar{z})$. Monatsh. f. Math., 72, 18-37 (1967).

[9] ---
Über die Darstellung von Lösungen einer partiellen Differentialgleichung mit N komplexen Variablen. Inst. f. Angew. Math., Univ. u. Techn. Hochsch. Graz, Ber. Nr. 70-4, 1-12 (1970).

[10] ---
Über Differentialgleichungen der Form $F(z,\bar{z})w_{z\bar{z}}-n(n+1)w=0$. Monatsh. f. Math., 75, 1-13 (1971).

[11] ---
Isolierte Singularitäten der Lösungen einer elliptischen

Differentialgleichung. Inst. f. Angew. Math., Univ. u. Techn. Hochsch. Graz, Ber. Nr. 71-1, 1-15 (1971).

[12] ---

Allgemeine Darstellungssätze bei einer Klasse partieller Differentialgleichungen gerader Ordnung. Monatsh. f. Math., 76, 193-213 (1972).

[13] ---

Eine Darstellung der allgemeinen Kugelfunktionen. Ber. d. Gesellsch. f. Math. u. Datenv., Bonn, Nr. 57, 5-11 (1972).

[14] ---

Differentialoperatoren bei partiellen Differentialgleichungen. Ber. d. Gesellsch. f. Math. u. Datenv., Bonn, Nr. 77, 7-17 (1973).

[15] ---

Differentialoperatoren bei einer Klasse verallgemeinerter Tricomi-Gleichungen. ZAMM, 54, 715-721 (1974).

[16] ---

Eine verallgemeinerte Darboux-Gleichung. Monatsh. f. Math., Teil I: 80, 1-11 (1975); Teil II: 80, 265-276 (1975).

[17] ---

Zur Lösungsdarstellung bei gewissen parabolischen Differentialgleichungen. Rend. Ist. Mat. Univ. Trieste, Vol. VII, Heft II, 116-127 (1975).

[18] ---

Polynomoperatoren bei Differentialgleichungen der Form $w_{z\bar{z}} + Aw_{\bar{z}} + Bw = 0$. Journ. Reine u. Angew. Math., 283/284, 364-369 (1976).

[19] ---

Zur Darstellung pseudo-analytischer Funktionen. Lecture Notes in Math., 561, 101-111 (1976).

[20] ---

Erzeugung und Darstellung von Lösungen eines verallgemeinerten Stokes-Beltrami-Systems. ZAMM, 57, 418-420 (1977).

[21] ---

Zur iterierten Gleichung der verallgemeinerten axial-symmetrischen Potentialtheorie. Akad. Wiss. SSSR, Steklov-Institut, Moskau, Festband zum 70. Geburtstag des Akademiemitgliedes I.N. Vekua: Komplexe Analysis und ihre Anwendungen, 45-54 (1978).

[22] ---

Bestimmung und Anwendung von Vekua-Resolventen. Monatsh.

f. Math., 85, 89-97 (1977).

[23] ---
On a differential equation in the theory of pseudo-holomorphic functions. J. Math. Soc. Japan, 30, 457-461 (1978).

[24] --- und H. Florian
Bergman-Operatoren mit Polynomerzeugenden. Research Notes in Math., 8, 85-93 (1976).

[25] --- und G. Jank
Differentialoperatoren bei einer inhomogenen elliptischen Differentialgleichung. Rend. Ist. Mat. Univ. Trieste, Vol. III, Heft II, 140-168 (1971).

[26] --- und E. Peschl
Ein allgemeiner Entwicklungssatz für die Lösungen der Differentialgleichung $(1+\varepsilon z\bar{z})^2 w_{z\bar{z}} + \varepsilon n(n+1)w = 0$ in der Nähe isolierter Singularitäten. Sitz.-Ber. Bayer. Akad. Wiss., math.-naturw. Kl., 113-146 (1965).

[27] --- und E. Peschl
Eindeutige Lösungen einer partiellen Differentialgleichung mit mehrdeutigen Erzeugenden. Arch. d. Math., XVIII, 285-289 (1967).

[28] --- und C. Rogers
Zur infinitesimalen Deformation von Flächen. Math.-stat. Sektion, Forsch.-Z. Graz, Ber. Nr. 31, 1-16 (1975).

[29] --- und St. Ruscheweyh
Ein Darstellungssatz für eine Klasse pseudoanalytischer Funktionen. Ber. d. Gesellsch. f. Math. u. Datenv., Bonn, Nr. 75, 3-15 (1973).

[30] --- und St. Ruscheweyh
Unterschiedliche Darstellungen für die Lösungen einer partiellen Differentialgleichung und deren Anwendungen. Math.-stat. Sektion, Forsch.-Z. Graz, Ber. Nr. 8, 1-20 (1974).

[31] Behnke, H. und P. Thullen
Theorie der Funktionen mehrerer komplexer Veränderlichen. Springer-Verlag, Berlin: 1934.

[32] Bergman, S.
Integral Operators in the Theory of Linear Partial Differential Equations. Erg. Math. Grenzgeb., Bd. 23, Springer-Verlag, Berlin-Göttingen-Heidelberg: 1961.

[33] Bers, L.
Theory of Pseudo-Analytic Functions. New York University, 1953.

[34] ---
An Outline of the Theory of Pseudo-Analytic Functions. Bull. Am. Math. Soc., 62, 291-331 (1956).

[35] Bitsadze, A.V.
On a Class of Nonlinear Partial Differential Equations. Lecture Notes in Math., 561, 10-16 (1976).

[36] --- und V.I. Paškovskiĭ
On the Theory of the Maxwell-Einstein Equations. Dokl. Akad. Nauk SSSR, Tom 216, Nr. 2, 762-774 (1974).

[37] Blum, E.K.
The Euler-Poisson-Darboux Equation in the Exceptional Cases. Proc. Am. Math. Soc., 5, 511-520 (1954).

[38] Burns, J.C.
The Iterated Equation of Generalized Axially Symmetric Potential Theory. Journ. Australian Math. Soc., Teil I-III: 7, 263-300 (1967), Teil IV: 9, 153-160 (1969), Teil V: 11, 129-141 (1970), Teil VI: 18, 318-327 (1974).

[39] Colton, D.
Cauchy's Problem for a Singular Parabolic Partial Differential Equation. J. Diff. Equ., 8, 250-257 (1970).

[40] Darboux, G.
Lecons sur la theorie general des surfaces. 2. Aufl., Gauthier-Villars, Paris: 1915.

[41] Diaz, J.B. und H.F. Weinberger
A Solution of the Singular Initial Value Problem for the Euler-Poisson-Darboux-Equation. Proc. Amer. Math. Soc., 4, 703-715 (1953).

[42] Erdélyi, A.
Singularities of Generalized Axially Symmetric Potentials. Comm. Pure and Appl. Math., 9, 403-414 (1956).

[43] Ernst, F.J.
New Formulation of the Axially Symmetric Gravitational Field Problem. Phys. Rev., 167, 1175-1178 (1968).

[44] ---
New Formulation of the Axially Symmetric Gravitational Field Problem, II. Phys. Rev., 168, 1415-1417 (1968).

[45] ---
Complex potential formulation of the axially symmetric gravitational field problem. Journ. Math. Phys., 15, 1409-1412 (1974).

[46] Elstrodt, J.
Die Resolvente zum Eigenwertproblem der automorphen Formen in der hyperbolischen Ebene. Teil I: Math. Ann., 203, 295-330 (1973), Teil II: Math. Z., 132, 99-134 (1973), Teil III: Math. Ann., 208, 99-132 (1974).

[47] Florian, H.
Integraloperatoren zur Lösung einer Klasse von Differentialgleichungen mit n Variablen. Inst. f. Angew. Math., Univ. u. Techn. Hochsch. Graz, Ber. Nr. 67-3, 1-13 (1967).

[48] --- und G. Jank
Polynomerzeugende bei einer Klasse von Differentialgleichungen mit zwei unabhängigen Variablen. Monatsh. Math., 75, 31-37 (1971).

[49] --- und R. Heersink
Über eine partielle Differentialgleichung mit p+2 Variablen und deren Zusammenhang mit den allgemeinen Kugelfunktionen. Manuscripta math., 12, 339-349 (1974).

[50] Friedlander, F.G.
Sound Pulses. Cambridge, England: Cambridge University Press 1958.

[51] --- und A.E. Heins
On the Singular Boundary Value Problem for the Euler-Darboux-Equation. J. Diff. Equ., 4, 460-491 (1968).

[52] --- und A.E. Heins
On the Representation Theorems of Poisson, Riemann and Volterra for the Euler-Poisson-Darboux-Equation. Arch. Ration. Mech. Anal., 33, 219-230 (1969).

[53] Friedman, A.
Partial Differential Equations of Parabolic Type. Prentice Hall, Englewood Cliffs: 1964.

[54] Gilbert, R.P.
On the Singularities of Generalized Axially Symmetric Potentials. Arch. Rat. Mech. Anal., 6, 171-176 (1960).

[55] ---
Function theoretic methods in partial differential equations. Academic Press, New York-London: 1969.

[56] ---
A method of ascent for solving boundary value problem. Bull. Amer. Math. Soc., Vol. 75, No. 6, 1286-1289 (1969).

[57] ---
The construction of solutions for boundary value problems

by function theoretic methods. SIAM J. Math. Anal., Vol. 1, No. 1, 96-114 (1970).

[58] Heersink, R.
Partial Differential Equations with Complex Variables. Matematica Balkanica, 4.42, 245-250 (1970).

[59] ---
Spezielle Operatoren zur Lösung partieller Differentialgleichungen. Inst. f. Angew. Math., Univ. u. Techn. Hochschule Graz, Ber. 72-2, 1-61 (1972).

[60] ---
Operatoren bei einer inhomogenen partiellen Differentialgleichung. Sitzungsber. Österr. Akad. Wiss., math.-naturw. Kl., Abt. II, 183, 361-372 (1974).

[61] ---
Characterisation of Certain Differential Operators in the Solution of Linear Differential Equations. Glasgow Math. Journ. Vol. 17, Part 2, 83-88 (1976).

[62] ---
Über Lösungsdarstellungen und funktionentheoretische Methoden bei elliptischen Differentialgleichungen. Math.-statistische Sektion, Forschungszentrum Graz, Bericht Nr. 67, 1-79 (1976).

[63] Huber, A.
On uniqueness of generalized axially symmetric potentials. Ann. of Math., 60, 351-385 (1954).

[64] ---
Some Results on Generalized Axially Symmetric Potential. Proc. Conf. Part. Diff. Equ., Maryland, 147-155 (1955).

[65] Jank, G.
Operatoren und partielle Differentialgleichungen 2. Ordnung. Inst. f. Angew. Math., Univ. u. Techn. Hochsch. Graz, Ber. Nr. 70-5, 1-37 (1970).

[66] ---
Integro-Differentialoperatoren bei partiellen Differentialgleichungen. Ber. d. Gesellsch. f. Math. u. Datenv., Bonn, Nr. 77, 91-95 (1973).

[67] ---
Integral- und Differentialoperatoren bei einer Differentialgleichung mit mehreren komplexen Veränderlichen. Periodica Mathematica Hungarica, Vol. 3-4 (3), 305-312 (1973).

[68] ---

Integral- und Differentialoperatoren bei einer Differentialgleichung mit mehreren komplexen Veränderlichen. Period. Math. Hungarica, Vol. 3, 305-312 (1973).

[69] ---

On integro-differential operators for partial differential equations. Research Notes in Math., 8, 158-171 (1976).

[70] --- und St. Ruscheweyh

Eine Bemerkung zur Darstellung gewisser pseudoanalytischer Funktionen. Ber. d. Gesellsch. f. Math. u. Datenv., Bonn, Nr. 75, 17-19 (1973).

[71] --- und K.J. Wirths

Generalized maximum principles in certain classes of pseudoanalytic functions. Research Notes in Math., 8, 63-67 (1976).

[72] Kamke, E.

Differentialgleichungen. Lösungsmethoden und Lösungen I. 6. verb. Aufl., Akad. Verlagsgesellschaft Geest u. Portig KG, Leipzig: 1959.

[73] Koohara, A.

Representation of pseudo-holomorphic functions of several complex variables. J. Math. Soc. Japan, 28, 257-277 (1976).

[74] Kracht, M.

Zur Existenz und Charakterisierung von Bergman-Operatoren. I: Der eingliedrige Lösungsansatz. Journ. Reine u. Angew. Math., 265, 202-220 (1974).

[75] ---

Über Bergman-Operatoren für lineare partielle Differentialgleichungen zweiter Ordnung. Habilitationsschrift, Düsseldorf 1974.

[76] --- und E. Kreyszig

Bergman-Operatoren mit Polynomen als Erzeugenden. Manuscripta math., 1, 369-376 (1969).

[77] --- und E. Kreyszig

Zur Konstruktion gewisser Integraloperatoren für partielle Differentialgleichungen. Manuscripta math., 17, Teil I: S. 79-103, Teil II: S. 171-186 (1975).

[78] --- und G. Schröder

Bergmansche Polynom-Erzeugende erster Art. Manuscripta math., 9, 333-355 (1973)

[79] Kreyszig, E.

Über zwei Klassen Bergman'scher Operatoren. Math. Nachr., 37, 197-202 (1968).

[80] ---
Bergman-Operatoren der Klasse P. Monatsh. Math., 74, 437-444 (1970).

[81] ---
On Bergman-Operators for Partial Differential Equations in Two Variables. Pac. J. Math., Vol. 36, Nr. 1, 201-208 (1971).

[82] Lamb, G.L. Jr.
π Pulse Propagation in a Lessless Amplifier. Phys. Letters, 29A, 507-508 (1969).

[83] ---
Higher Conversation Laws in Ultrashort Optical Pulse Propagation. Phys. Letters, 32A, 251-252 (1970).

[85] ---
Analytic Descriptions of Ultrashort Optical Pulse Propagation in a Resonant Medium. Rev. Mod. Phys., 43, 99-124 (1971).

[85] Lanckau, E.
Eine Anwendung der Bergman'schen Operatorenmethode auf Profilströmungen im Unterschall. Wiss. Z. Techn. Hochsch. Dresden, 8, 200-207 (1958/59).

[86] Maaß, H.
Über eine neue Art von nichtanalytischen automorphen Funktionen und die Bestimmung Dirichlet'scher Reihen durch Funktionalgleichungen. Math. Ann., 121, 141-183 (1949).

[87] Magnus, W., F. Oberhettinger und R.P. Soni
Formulas and Theorems for the Special Functions of Mathematical Physics. Third enlarged edition. Springer-Verlag, Berlin-Heidelberg-New York: 1966.

[88] Miranda, C.
Equazioni alle derivate parziali di tipo ellitico. Erg. Math. Grenzgeb., Neue Folge, Heft 2, Springer-Verlag, Berlin-Göttingen-Heidelberg: 1955.

[89] Müller, Cl.
Spherical Harmonics. Lecture Notes in Math., 17, Springer-Verlag, Berlin-Heidelberg-New York: 1966.

[90] Payne, L.E.
Representation Formulas for Solutions of a Class of Partial Differential Equations. J. Math. and Phys., 38, 145-149 (1959).

[91] Power, G., C. Rogers und R.A. Osborn
Bäcklund and Generalized Legendre Transformations in Gas-

dynamics. Z. Angew. Math. Mech., 49, 333-340 (1969).

[92] Püngel, J.
Zur Darstellung von Lösungen partieller Differentialgleichungen. Math.-statistische Sektion, Forschungszentrum Graz, Bericht Nr. 72, 1-10 (1977).

[93] ---
Lineare Abbildungen zwischen Lösungsmengen partieller Differentialgleichungen zweiter Ordnung im Komplexen. Math.-statistische Sektion, Forschungszentrum Graz, Bericht Nr. 91, 1-81 (1978).

[94] Reich, L.
Über multiplikative und algebraische verzweigte Lösungen der Differentialgleichungen $(1+\varepsilon z\bar{z})^2 w_{z\bar{z}} + \varepsilon n(n+1)w = 0$. Ber. d. Gesellsch. f. Math. u. Datenv., Bonn, Nr. 57, 13-28 (1972).

[95] Roelcke, W.
Über die Wellengleichung bei Grenzkreisgruppen erster Art. Sitz.-Ber. Heidelberger Akad. Wissen., Math.-natw. Kl., 1953-1955, 159-267 (1956).

[96] Rogers, C.
Application of Bäcklund Transformations in Aligned Magnetogasdynamics. Acta Phys. Austriaca, 31, 80-88 (1970).

[97] --- und J.G. Kingston
On Certain Matrix Transformations of the Stokes-Beltrami Equations. Tensor, N.S., 22, 269-273 (1971).

[98] --- und J.G. Kingston
Baecklund Transformations with Inversion Applied to the Stokes-Beltrami Equations. Tensor, N.S., 24, 322-328 (1972).

[99] Ruscheweyh, St.
Über den Rand des Einheitskreises fortsetzbare Lösungen der Differentialgleichung von Peschl und Bauer. Ber. d. Gesellsch. f. Math. u. Datenv., Bonn, Nr. 57, 29-36 (1972).

[100] Scott, A.C.
Propagation of Magnetic Flux in a Long Josephson Junction. Nuovo Cimento, 69B, 241-261 (1970).

[101] Vekua, I.N.
Verallgemeinerte analytische Funktionen. Akademie-Verlag, Berlin: 1963.

[102] ---
New Methods for Solving Elliptic Equations. North-Holland Publ. Co., Amsterdam: 1968.

[103] Walter, W.
Über die Euler-Poisson-Darboux-Gleichung. Math. Z., 67, 361-376 (1957).

[104] Warnecke, G.
Über die Darstellungen von Lösungen der partiellen Differentialgleichung $(1+\delta z\bar{z})^2 w_{z\bar{z}} = \delta - \varepsilon e^{2w}$. Bonner Math. Schriften, Nr. 34, 1-75 (1968).

[105] Watzlawek, W.
Über lineare partielle Differentialgleichungen zweiter Ordnung mit Fundamentalsystemen. Journ. Reine u. Angew. Math., 247, 69-74 (1971).

[106] ---
Über Zusammenhänge zwischen Fundamentalsystemen, Riemann-Funktion und Bergman-Operatoren. Journ. Reine u. Angew. Math., 251, 200-211 (1971).

[107] ---
Über lineare partielle Differentialgleichungen zweiter Ordnung mit Bergman-Operatoren der Klasse P. Monatsh. f. Math., 76, 356-369 (1972).

[108] ---
Hyperbolische und parabolische Differentialgleichungen der Klasse P. Ber. d. Gesellsch. f. Math. u. Datenv., Bonn, Nr. 77, 147-179 (1973).

[109] Weinacht, R.J.
Fundamental Solutions for a Class of Singular Equations. Contributions to Differential Equations, Vol. III, 43-55 (1964).

[110] Weinstein, A.
Generalized Axially Symmetric Potential Theory. Bull. Am. Math. Soc., 59, 20-38 (1953).

[111] ---
The Singular Solution and the Cauchy Problem for Generalized Tricomi Equations. Comm. Pure Appl. Math., Vol. VII, 105-116 (1954).

[112] ---
On a Class of Partial Differential Equations of Even Orders. Ann. Mat. Pura Appl., 39, 245-254 (1955).

[113] ---
Singular Partial Differential Equations and Their Applications: Fluid Dynamics and Applied Mathematics (Proceedings of a Symposium, held at the University of Maryland, 1961) 22-49. Gordon and Breach, New York: 1962.

SUBJECT INDEX

PART II

Stephan Ruscheweyh

On the Function Theory of the Peschl-Bauer Equation

INTRODUCTION

K.W. Bauer's discovery of the differential operator representation for the solutions of

$$(0.1) \qquad (1+\varepsilon z\bar{z})^2 w_{z\bar{z}} + \varepsilon n(n+1)w = 0, \qquad \varepsilon = \pm 1, \qquad n \in \mathbb{N},$$

(implicitly known already to G. Darboux) has been the germ of the theory presented in the preceding report.
The importance of (0.1) is obvious from the fact that

$$(1+\varepsilon z\bar{z})^2 \frac{\partial^2}{\partial z \partial \bar{z}}$$

is the Laplace operator of the spherical ($\varepsilon = 1$) and the hyperbolical ($\varepsilon = -1$) geometry respectively. The particular solutions

$$(0.2) \qquad P_n\left(\frac{1-\varepsilon z\bar{z}}{1+\varepsilon z\bar{z}}\right), \qquad Q_n\left(\frac{1-\varepsilon z\bar{z}}{1+\varepsilon z\bar{z}}\right),$$

where P_n, Q_n denote the Legendre functions of the first and second kinds show that (0.1) is related to the special functions of mathematical physics. It is also distinguished by the fact that it is one of the very few equations for which the Riemann function of Vekua's theory is known. Furthermore, (0.1) is closely connected with a number of important and thoroughly studied equations. For instance, if we map the unit disc $E = \{z \mid |z| < 1\}$ conformally onto the upper half plane ($\eta = i(1+z)/(1-z)$) then (0.1) with $\varepsilon = -1$ transforms into

$$(0.3) \qquad (\eta-\bar{\eta})^2 W_{\eta\bar{\eta}} + n(n+1)W = 0,$$

which is the equation of certain Eisenstein series. Nonanalytic automorphic forms as solutions of (0.3) are frequently investigated (see e.g. [40], [49], [21]). The further transformation $U(\eta)=(\eta-\bar{\eta})^{-n-1}W(\eta)$ gives

$$(0.4) \qquad (\eta-\bar{\eta})U_{\eta\bar{\eta}} - (n+1)(U_\eta - U_{\bar{\eta}}) = 0.$$

This equation is known as the GASP-equation; its rich theory has been surveyed by A. Weinstein [66]. We also mention the related chapter in R.P. Gilbert's book [24] and A. Huber's papers [30,31].
The present report deals exclusively with those results on (0.1) which - directly or indirectly - arose from Bauer's representation theorem. This is why we do not refer to the Maass-Roelcke theory on automorphic forms, for instance, but discuss the representation of automorphic solutions of (0.1) with Bauer operators and their interesting connection with the Eichler integrals.
The function theoretic approach to (0.1) was initiated in 1965 by K. W. Bauer [4] and since then quite a number of mathematicians has participated in its development. The present author found this field of research always very stimulating, even for related studies in classical function theory.
The results obtained so far cover a number of different topics. Large gaps, however, are still remaining. While e.g. in the 'geometric' theory a good amount of information has been collected (although the local behaviour is not yet sufficiently classified) there are other fields like 'entire functions' where no deeper results are available. We believe that recently developed methods on value distribution of entire functions and their derivatives (cf. G. Frank [22]) may apply to the entire solutions of (0.1) and may produce progress also in that direction. So it is one of the main intentions of this report to create interest in such problems and to encourage other mathematicians to work in this new field.
This paper contains published and unpublished results. Therefore some of the theorems appear with detailed proofs, others without any. However, we always tried to outline the basic ideas so that not much reference to original papers should be necessary.
I wish to dedicate this work to Prof. K.W. Bauer, who was my teacher and patient guide for many years.

CHAPTER 1

Structure of Solutions

1.1. Let $G \subset G_\varepsilon$ be a domain where G_ε is the unit disc E ($\varepsilon = -1$) and the Riemannian sphere ($\varepsilon = 1$) respectively. $\Omega_n(G)$ is the set of two times continuously differentiable solutions of (0.1) in G. In particular, $\Omega_o(G)$ denotes the set of complex valued harmonic functions in G. For certain domains G a representation of the functions in $\Omega_n(G)$ has been given by I.N. Vekua [64]: if $G \subset E$ is starlike w.r.t. $0 \in G$ then the mapping $T_n: \Omega_o(G) \to \Omega_n(G)$ with

$$(1.1) \qquad (T_n u)(z) = u(z) - \int_0^1 u(tz) \frac{\partial}{\partial t} P_n\left(\frac{1+(2t-1)\varepsilon z\bar{z}}{1+\varepsilon z\bar{z}}\right)dt, \quad z \in G,$$

is bijective [64, (14.11)].
If u is harmonic in G then

$$h(z) = \int_0^1 \frac{(1-t)^{n-1}}{(n-1)!} u(tz)dt, \qquad z \in G,$$

has the same property and can be decomposed into $h = h_1 + \overline{h_2}$, where h_1, h_2 are analytic in G. Using the expansion

$$P_n(x) = \sum_{k=0}^{n} \frac{(n+k)!}{2^k (k!)^2 (n-k)!} (x-1)^k$$

we obtain from (1.1)

$$(1.2) \qquad T_n u = E_n(z^n h_1) + \overline{E_n(z^n h_2)}$$

where

$$(1.3) \qquad E_n = \sum_{k=0}^{n} \frac{(n+k)!}{k!(n-k)!} \left(\frac{-\varepsilon\bar{z}}{1+\varepsilon z\bar{z}}\right)^k \frac{d^{n-k}}{dz^{n-k}}$$

is Bauer's operator. Bauer [3] [1] has shown that (1.2) holds for more general domains. To state his result in a short form we make use of the following identity [2]:

$$E_n(z^n h_1) + \overline{E_n(z^n h_2)} = \frac{(-\varepsilon)^n}{n!}(1+\varepsilon z\bar{z})^{n+1}\Delta^n\left(\frac{h_1(z)+\overline{h_2(z)}}{1+\varepsilon z\bar{z}}\right)$$

where h_1, h_2 are analytic in G.

<u>THEOREM 1.1.</u> Let $G \subset G_\varepsilon$ be a simply connected domain. Then the operator

$$(S_n h)(z) = \frac{(-\varepsilon)^n}{(n!)^2}(1+\varepsilon z\bar{z})^{n+1}\Delta^n\left(\frac{h(z)}{1+\varepsilon z\bar{z}}\right) \tag{1.4}$$

generates a surjective mapping S_n: $\Omega_o(G) \to \Omega_n(G)$.

<u>REMARKS.</u> 1) If $0 \in G$ then this mapping is even bijective. Otherwise, for $w = S_n h$, the most general harmonic function g with $w = S_n g$ is given by

$$g(z) = h(z) + P(z) - P(-\varepsilon/\bar{z}),$$

where P is an arbitrary function of the form $P(z) = z^{-n}Q(z)$, Q polynomial of degree $\leq 2n$.

2) The particular solutions mentioned in (0.2) have the representation

$$P_n\left(\frac{1-\varepsilon z\bar{z}}{1+\varepsilon z\bar{z}}\right) = S_n 1, \quad Q_n\left(\frac{1-\varepsilon z\bar{z}}{1+\varepsilon z\bar{z}}\right) = S_n(-\log|z|). \tag{1.5}$$

Furthermore, for $k \in \mathbb{N}$,

$$S_n(z^k) = z^k\binom{n+k}{n}\,{}_2F_1\left(-n,n+1;k+1;\frac{\varepsilon z\bar{z}}{1+\varepsilon z\bar{z}}\right), \quad k \in \mathbb{N}. \tag{1.6}$$

---

[1] Cf. R. Heersink [27], L. Reich [48]. M. Kracht and E. Kreyszig [36] have shown that Bauer's representation - for special domains - can also be obtained from Bergman's integral operator for (0.1).

[2] $\Delta^0 = \mathrm{id}$, $\Delta^1 = \frac{\partial^2}{\partial z\partial\bar{z}}$, $\Delta^n = \Delta(\Delta^{n-1})$.

These functions are related to the associated Legendre functions, cf. [1; (15.4.17)].

Particular attention will be paid to the following subclasses of $\Omega_n(G)$:

$$\Omega_n'(G) = \{S_n h \mid h \text{ analytic in } G\},$$

$$\Omega_n''(G) = \{S_n(h+\bar{P}) \mid h \text{ analytic in } G,\ P \text{ polynomial, } \deg P \leq n\}.$$

Since for polynomials P with deg $P \leq n$ the relation

$$S_n(P(z)) = S_n(P(-\varepsilon/\bar{z})) \tag{1.7}$$

holds we conclude that $\Omega_n''(G)$ coincides with the class

$$\{E_n g \mid g \text{ analytic in } G\}$$

previously considered by Bauer. Further, for domains G with $0 \in G$ we have $\Omega_n'(G) = \Omega_n''(G)$.

In many cases it is preferable to deal with sets of functions which contain all constants. For $\Omega_n(G)$ this is not valid but sometimes we circumvent this problem by considering the class $\Lambda_n(G)$ of functions[1)]

$$v(z) = w(z)/(S_n 1)(z), \quad w \in \Omega_n(G),$$

which has the desired property. However, the following restriction has to be observed: in case $\varepsilon = 1$ we define $\Lambda_n(G)$ only for domains $G \subset E_{t_n}$[2)], where

$$t_n = \text{least positive zero of } (S_n 1)(t). \tag{1.8}$$

In such domains $\Lambda_n(G)$ is the set of solutions of the equation

$$v_{z\bar{z}} + \left[\frac{1}{2|z|}\frac{d}{d|z|}\log(S_n 1)\right](zv_z + \bar{z}v_{\bar{z}}) = 0, \tag{1.9}$$

---

1) $\Lambda_n'(G)$ is defined correspondingly.

2) $E_R = \{z \mid |z| < R\}$.

and from our previous observations we conclude that for $0 \in G \subset E_{t_n}$ the operator

$$(1.10) \qquad (F_n h)(z) = \frac{\Delta^n[h(z)(1+\varepsilon z\bar{z})^{-1}]}{\Delta^n[(1+\varepsilon z\bar{z})^{-1}]}$$

generates a bijective mapping F_n: $\Omega_o(G) \to \Lambda_n(G)$.

1.2. Equation (0.1) is invariant against rotations of the sphere S^2 ($\varepsilon = 1$) and conformal automorphisms of E ($\varepsilon = -1$) respectively. In fact, for

$$(1.11) \qquad \eta = \eta(z) = e^{i\alpha} \frac{z+a}{1-\varepsilon\bar{a}z}, \quad a \in G_\varepsilon, \quad \alpha \in \mathbb{R},$$

and $w(\eta) \in \Omega_n(G)$ we have $W(z) = w(\eta(z)) \in \Omega_n(\eta^{-1}(G))$. I.N. Vekua while determining the Riemann function for (0.1) made good use of this property.
The operator E_n behaves as follows: for g analytic in $G \subset G_\varepsilon$ we have

$$(1.12) \qquad (E_n g)(\eta) = (E_n h)(z), \quad h(z) = [\eta'(z)]^{-n} g(\eta(z)).$$

Thus $\Omega_n''(G)$ is 'invariant' with respect to such transformations:

$$w(\eta) \in \Omega_n''(G) \iff w(\eta(z)) \in \Omega_n''(\eta^{-1}(G)).$$

This valuable property fails to hold for $\Omega_n'(G)$ and $\Lambda_n(G)$.
(1.12) can be used to derive the following useful representation for E_n:

$$(1.13) \qquad (E_n g)(z) = \frac{d^n}{dx^n}\left[g\left(\frac{x+z}{1-\varepsilon x\bar{z}}\right) \frac{(1-\varepsilon x\bar{z})^{2n}}{(1+\varepsilon z\bar{z})^n}\right]\Bigg|_{x=0}$$

wherever g is analytic.

1.3. Let $z = re^{i\varphi}$. Then, if $w \in \Omega_n(G)$, the same holds for w_φ, in particular

$$\frac{\partial}{\partial\varphi} S_n(h) = S_n\left(\frac{\partial}{\partial\varphi} h\right), \quad \frac{\partial}{\partial\varphi} F_n(h) = F_n\left(\frac{\partial}{\partial\varphi} h\right).$$

The corresponding property for differentiation w.r.t. r is not valid. (1.4), however, leads to the following recursions:

$$S_{n+1}(h) = \frac{1-\varepsilon r^2}{1+\varepsilon r^2} S_n(h) + \frac{r}{n+1} \frac{\partial}{\partial r} S_n(h) , \tag{1.14}$$

$$F_{n+1}(h) = F_n(h) + \frac{S_n(1)}{(n+1)S_{n+1}(1)} r \frac{\partial}{\partial r} F_n(h), \tag{1.15}$$

for harmonic h. For

$$g = \left[\left(r \frac{\partial}{\partial r}\right)^2 - n^2\right] h$$

we obtain

$$S_{n-1}(g) = -n^2 \frac{1-\varepsilon r^2}{1+\varepsilon r^2} S_n(h) + nr \frac{\partial}{\partial r} S_n(h) . \tag{1.16}$$

Further similar relations can be found in Bauer [8].
For $h \equiv 1$ (1.14) and (1.16) are well known recursions for Legendre polynomials:

$$P_{n+1}(x) = xP_n(x) + \frac{x^2-1}{n+1} P_n'(x) ,$$

$$P_{n-1}(x) = xP_n(x) - \frac{x^2-1}{n} P_n'(x) .$$

$h = \log |z|$ gives the corresponding relations for Legendre functions of the second kind.
Another important structural property (multiplication theorem) for (0.1) follows from an interesting but apparently almost unknown generalization of Leibniz' rule (J.J. Walker [65], cf.[53]): for f,g,h n-times continuously differentiable in (a,b), $f(x) \neq 0$ in (a,b), we have

$$(gh)^{(n)} = gh^{(n)} + \sum_{k=1}^{n} \binom{n}{k} (g'f^{-k})^{(k-1)}(hf^k)^{(n-k)}.$$

An application of this formula to (1.4) gives

$$(1.17)\qquad S_n(gh) = gS_n(h) + \sum_{k=0}^{n-1} \frac{1}{n-k} S_k(h)S_{n-k-1}(zg')$$

where g,h are analytic in a domain $G \subset G_\varepsilon$. Also this formula generalizes well known relations between Legendre functions. E.g. putting h = 1, g = - log z and taking real parts gives Schläfli's representation of Legendre functions of the second kind:

$$Q_n(X) = -P_n(X)\log|z| - \sum_{k=0}^{n-1} \frac{1}{n-k} P_k(X)P_{n-k-1}(X),$$

$X = (1-\varepsilon z\bar{z})/(1+\varepsilon z\bar{z})$. Other choices of h, g lead to analogous results for associate Legendre functions.

1.4. Equations (0.1) with $\varepsilon = -1$ and (1.9) are of the form $D(w) = 0$ where

$$D(w) = w_{z\bar{z}} + a(z,\bar{z})(zw_z + \bar{z}w_{\bar{z}}) + b(z,\bar{z})w\ ,$$

a, b real analytic, $b \leq 0$. If $D(w) = 0$ in a domain with $w \neq 0$ then for $p > 0$:

$$(1.18)\qquad D(|w|^p) = |w|^p\left[\frac{p(p-1)}{4}\left|\frac{w_z}{w} + \frac{\bar{w}_{\bar{z}}}{\bar{w}}\right|^2 + \frac{p}{4}\left|\frac{w_z}{w} - \frac{\bar{w}_{\bar{z}}}{\bar{w}}\right|^2 - (p-1)b\right]\ .$$

For $p \geq 1$ this implies $D(|w|^p) \geq 0$ in G and by E. Hopf's maximum principle (cf. [47]) we can conclude that $|w|$ cannot assume its maximum in G without being constant. In the latter case, however, we have w = const.. In fact, let $w = u + iv$ and $u^2 + v^2 = C$. Then by (1.18)

$$D(u^2) \geq 0,\quad D(C-u^2) \geq 0,\quad D(C) \leq 0,$$

which implies $D(u^2) = |u_z|^2 + |u_{\bar{z}}|^2 - bu^2 = 0$. Thus u^2 = const. and w = const..

In case $\varepsilon = 1$ no general maximum principle exists. A week form, however, can be established also for these functions, cf. section 3.3.

Let w fulfil $D(w) = 0$ with $b \equiv 0$. Then, for all $\alpha, \beta \in \mathbb{C}$ and $v = \mathrm{Re}(\alpha w+\beta)$, the maximum principle is valid. This can be used to obtain better information about the range of w. For equation (1.9) we obtain:

<u>THEOREM 1.2.</u> Let $w \in \Lambda_n(G)$ be continuous in $\bar{G}$. Then [1)]

$$w(G) \subset \overset{\bullet}{\text{clco}}(w(\partial G)). \tag{1.19}$$

Many 'geometrical' results in the subsequent chapters are based on Theorem 1.2.

1.5. Let $z = re^{i\varphi}$. With any function $h(z)$ harmonic in $E_R \subset G_\varepsilon$

$$h(z) = \sum_{k=-\infty}^{\infty} a_k r^{|k|} e^{ik\varphi}$$

we associate the functions [2)]

$$s_n(z,t;h) = \sum_{k=-\infty}^{\infty} a_k r^{|k|} s_n(t^{|k|}) e^{ik\varphi}, \quad 0 < t < R. \tag{1.20}$$

Since

$$\lim_{k\to\infty} (s_n(t^k))^{1/k} = t$$

we see that $s_n(z,t;h)$ is harmonic in $|z| < R/t$. Furthermore, the linearity of S_n implies

$$(S_n h)(z) = s_n(e^{i\varphi}, r; h), \quad z \in E_R. \tag{1.21}$$

In a similar manner we define f_n which satisfy

$$(F_n h)(z) = f_n(e^{i\varphi}, r; h), \quad z \in E_R, \tag{1.22}$$

[1)] $\overset{\bullet}{\text{clco}}$ is the interior of the closed convex hull of a set, taken w.r.t. the relative topology.

[2)] $s_n(t^k) = s_n(z^k)\big|_{z=t}$.

where $R < t_n$ if $\varepsilon = 1$. For an analytic function g in E_R the following result is a consequence of Theorem 1.2, (1.22) and the argument principle.

<u>THEOREM 1.3.</u> For $0 < r < R \quad (< t_n)$

$$f_n(E,r;g) \subset (F_n g)(E_r) \subset \dot{\mathrm{clco}}(f_n(E,r;g)). \tag{1.23}$$

In order to extend this result to the important limiting case R = 1 ($\varepsilon = -1$), we need asymptotic expansions ($t \to 1-0$):

$$\frac{(n!)^2 2^n}{(2n)!}(1-t)^n s_n(t^k) = 1 - \frac{n}{2}(1-t) + \frac{2n^3-3n^2+3n-4k^2}{16n-8}(1-t)^2 + \\ + O((1-t)^3). \tag{1.24}$$

For a harmonic function h in E this implies:

$$s_n(z,t;h) = \frac{(2n)!}{(n!)^2 2^n} \frac{1 - \frac{n}{2}(1-t)}{(1-t)^n} h(z) + O((1-t)^{2-n}), \tag{1.25}$$

$$f_n(z,t;h) = h(z) - \frac{(1-t)^2}{4n-2} \left(z \frac{\partial}{\partial |z|}\right)^2 h(z) + O((1-t)^3). \tag{1.26}$$

These expansions hold uniformly in compacta of E.

<u>THEOREM 1.4.</u> Let $\varepsilon = -1$, g analytic in E. Then

$$\overline{g(E)} \subset \overline{(F_n g)(E)} \subset \mathrm{clco}(g(E)) . \tag{1.27}$$

The Hadamard product

$$(g*h)(z) = \sum_{k=0}^{\infty} a_k b_k z^k$$

of two functions

$$g(z) = \sum_{k=0}^{\infty} a_k z^k \in \Omega_o'(E_{R_1}), \quad h(z) = \sum_{k=0}^{\infty} b_k z^k \in \Omega_o'(E_{R_2})$$

is analytic in $|z| < R_1 R_2$. With $g_o(z) = 1/(1-z)$ we have $g*g_o = g$ for every analytic g in E_R. This can be used to establish

$$s_n(z,t;g) = g*s_n(z,t;g_o), \quad f_n(z,t;g) = g*f_n(z,t;g_o), \tag{1.28}$$

which enables us to apply known theorems on Hadamard products of analytic functions to solutions of (0.1) and (1.9).

1.6. M.B. Balk introduced 'polyanalytic' functions

$$w = \sum_{k=0}^{n} g_k(z)\bar{z}^k, \quad g_k \text{ analytic in } G, \tag{1.29}$$

and established many function theoretic results for them (cf. the survey article [2]). Obviously there is a formal similarity with Bauer's operator

$$E_n g = \sum_{k=0}^{n} \frac{(n+k)!}{k!(n-k)!} g^{(n-k)}(z)\left(\frac{-\varepsilon\bar{z}}{1+\varepsilon z\bar{z}}\right)^k, \quad g \text{ analytic in } G.$$

In fact, some of Balk's results remain valid for the functions in $\Omega_n''(G)$.
Even more important is Krajkiewicz' [37] generalization of polyanalytic functions: he calls w 'multianalytic' in a punctured neighborhood $U_p(0)$ of the origin if

$$w = \sum_{k=-m}^{\infty} g_k(z)\bar{z}^k, \quad g_k \text{ analytic in } U_p(0).$$

Because of the following theorem his results directly apply to our case.

<u>THEOREM 1.5.</u> Let g, h be analytic in $U_p(0)$ where h has no essential singularity at $z = 0$. Then $S_n(g+\bar{h})$ is multianalytic in $U_p(0)$. In particular, every function in $\Omega_n(U(0))$ is multianalytic.

The proof is immediate from the expansion

$$(1.30) \qquad E_n g = \sum_{k=0}^{\infty} [(-\varepsilon)^k \sum_{j=0}^{\min(k,n)} \frac{(n+j)!}{j!(n-j)!} (-z)^{k-j} g^{(n-j)}(z)] \bar{z}^k$$

valid for g analytic in $U_p(0)$.

1.7. The following observation is useful for some applications: there exist rational functions $B_{nk}^{\varepsilon}(t)$ such that

$$(1.31) \qquad (S_n g)(z) = \sum_{k=0}^{n} (1+\varepsilon z\bar{z})^k g^{(k)}(z) z^k B_{nk}^{\varepsilon}(|z|)$$

holds for g analytic in G. In particular $B_{no}^{\varepsilon}(t) = (S_n 1)(t)$ and for $k = 1, \ldots, n$ the functions $B_{nk}^{\varepsilon}(t)/B_{no}^{\varepsilon}(t)$ have no singularities in $0 \le t \le 1$ $(\varepsilon = -1)$ and $0 \le t < t_n$ $(\varepsilon = 1)$.

CHAPTER 2

Dirichlet Problems for Circles

2.1. The Dirichlet problem (D) for circles E_R and prescribed continuous boundary values has in case $\varepsilon = -1$, $R < 1$, a uniquely determined solution (cf.[34]): if

$$f \sim \operatorname{Re} \sum_{k=0}^{\infty} a_k e^{ik\varphi}$$

is the Fourier expansion of the boundary function and

$$h(z) = \operatorname{Re} \sum_{k=0}^{\infty} \frac{a_k}{S_n(R^k)} z^k$$

then h is harmonic in E_R and $S_n h$ is the solution of (D) for f and E_R.

In case $\varepsilon = 1$ the situation is different since the coefficient of w in (0.1) becomes positive so that the general methods ensure existence and uniqueness of solutions of (D) only for $R < t_n$. For $R = 1$ M.P. Ganin [23] has shown that the solutions of (D) may not be uniquely determined or solutions may not even exist. Bauer [5] studied this problem for arbitrary R and proved unique solvability for all boundary functions and all but finitely many circles (for n fixed). His method was to establish the Green's function using his E_n operator. Only recently, R. Heersink [28] treated (D), including the exceptional circles, for höldercontinuous boundary values by the means of Floquet's theory. In the present chapter we shall apply the Banach-Steinhaus theorem to obtain a simple and complete solution of (D) for $\varepsilon = 1$.
We start with the uniqueness problem which has to be considered only for the boundary function $f \equiv 0$. Let

$$w = \operatorname{Re} \sum_{k=0}^{\infty} a_k S_n(z^k), \qquad a_o \in \mathbb{R},$$

be a solution of (D) with vanishing boundary values. Then for $r < R$

$$\frac{1}{2\pi} \int_0^{2\pi} w(re^{i\varphi}) e^{-ik\varphi} d\varphi = \begin{cases} a_k S_n(r^k)/2, & k \in \mathbb{N}, \\ a_o S_n(r^o), & k = 0, \end{cases}$$

and with $r \to R$: $a_k S_n(R^k) = 0$, $k \in \mathbb{N}_o$ [1]. This shows that a solution of (D) for E_R is unique if $S_n(R^k) \neq 0$, $k \in \mathbb{N}_o$. From

$$(2.1) \qquad S_n(R^k) = \left(\frac{R}{1+R^2}\right)^k \binom{k+n}{n} {}_2F_1(k+n+1,k-n;k+1;R^2/(1+R^2))$$

(cf. (1.6) and [1], Formula 15.3.3) we see $S_n(R^k) \neq 0$, $R > 0$, $k \geq n$. On the other hand we have from (1.6)

$$S_n(R^k) = \frac{2^k (n-k)!}{n!} \left(\frac{R}{1+R^2}\right)^k P_n^{(k)}\left(\frac{1-R^2}{1+R^2}\right), \qquad 0 \leq k < n,$$

such that the exceptional radii are the zeros of

$$P_n^{(k)}\left(\frac{1-R^2}{1+R^2}\right), \qquad k = 0,1, \ldots, n-1.$$

These admit the eigensolutions $\mathrm{Re}[\alpha S_n(z^k)]$, $\alpha \in \mathbb{C}$. We note that Ganin's case $R = 1$ is exceptional for every n.
Now we turn to the existence problem. For $0 < r < R$ let

$$K_R(\varphi,r) = a_o(r) + 2\sum_{k=1}^{\infty} a_k(r)\cos(k\varphi)$$

where

$$a_k(r) = \begin{cases} S_n(r^k)/S_n(R^k), & \text{for } S_n(R^k) \neq 0, \\ 1, & \text{otherwise.} \end{cases}$$

$K_R(\varphi,r)$ is continuous in φ and we shall show

$$(2.2) \qquad \sup_{r<R} \int_0^{2\pi} |K_R(\varphi,r)|\, d\varphi < \infty.$$

In fact, for fixed $n \in \mathbb{N}$, $v \in (0,1)$ the function $M(k,u)$ in

---

1) $\mathbb{N}_o = \mathbb{N} \cup \{0\}$.

$$(2.3)\qquad \frac{{}_2F_1(-n,n+1;k;u)}{{}_2F_1(-n,n+1;k,v)} = 1 + (v-u)\left[\frac{n(n+1)}{k} + \frac{M(k,u)}{k^2}\right]$$

is uniformly bounded in $[0,\infty]\times[0,v]$. This implies the existence of $k_o(n,v)$ such that the functions (2.3), for $k \geq k_o$, are positive, monotonically decreasing and convex in k. Hence, by (1.6) we conclude that $s_n(r^k)/s_n(R^k)$ are positive, monotonically decreasing and convex for $k \geq k_o(n,R^2/(1+R^2))$. This enables us to construct numbers $b_k > 0$, $k = 0, \ldots, k_o-1$, such that for

$$c_k(r) = \begin{cases} b_k, & 0 \leq k < k_o, \\ a_k(r), & k \geq k_o, \end{cases}$$

and $0 \leq r < R$ we have

$$c_k(r) - c_{k+1}(r) \geq c_{k+1}(r) - c_{k+2}(r) \geq 0, \qquad k \in \mathbb{N}_o.$$

An application of a well known result of W. Rogosinski gives

$$g_r(\varphi) = c_o(r) + 2\sum_{k=1}^{\infty} c_k(r)\cos(k\varphi) \geq 0, \qquad \varphi \in \mathbb{R},$$

and thus

$$\int_0^{2\pi} |K_R(\varphi,r)|d\varphi$$

$$= \int_0^{2\pi} |a_o(r)-b_o + 2\sum_{k=1}^{k_o-1} (a_k(r)-b_k)\cos(k\varphi) + g_r(\varphi)|d\varphi$$

$$\leq 4\pi \sum_{k=0}^{k_o-1} (a_k(r) + b_k).$$

Furthermore we have

$$(2.4)\qquad \lim_{r \to R-0} a_k(r) = 1, \qquad k \in \mathbb{N}_o .$$

(2.2) and (2.4) are the assumptions of a Corollary to the Banach-Steinhaus theorem (cf. [16], Th. 1.3.5) and imply for every f continuous on

∂E_R

$$\lim_{r \to R-0} \sup_{|\sigma|\le\pi} \left| \frac{1}{2\pi} \int_0^{2\pi} f(Re^{i\varphi})K_R(\sigma-\varphi,r)d\varphi - f(Re^{i\sigma}) \right| = 0.$$

For every non-exceptional R we conclude that (D) for E_R and arbitrary continuous f is solved by

$$(2.5) \qquad w(z) = \frac{1}{2\pi} \int_0^{2\pi} f(Re^{i\varphi})K_R(\sigma-\varphi,r)d\varphi, \qquad z = re^{i\sigma} .$$

If R is exceptional with the eigensolutions $Re[\alpha_j S_n(z^{k_j})]$, $j = 1, \ldots, s$, it is clear from the previous remarks that (D) is unsolvable if

$$\int_0^{2\pi} f(Re^{i\varphi})e^{-ik_j\varphi} d\varphi \neq 0$$

for at least one $j \in \{1, \ldots, s\}$. Otherwise (2.5) solves (D) also in this case and the general solution is given by

$$w(z) = \frac{1}{2\pi} \int_0^{2\pi} f(Re^{i\varphi})K_R(\sigma-\varphi,r)d\varphi + \sum_{j=1}^{s} Re[\alpha_j S_n(z^{k_j})]$$

with $\alpha_j \in \mathbb{C}$, $j = 1, \ldots, s$, arbitrary. We have proved the following theorem:

<u>THEOREM 2.1.</u> ($\varepsilon = 1$). Let $R > 0$ and for $s \in \mathbb{N}_0$, $j = 1, \ldots, s$, assume

$$P_n^{(k_j)}\left(\frac{1-R^2}{1+R^2}\right) = 0.$$

For

$$f \sim Re \sum_{k=0}^{\infty} a_k e^{ik\varphi} , \qquad a_0 \in \mathbb{R},$$

assume

$$(2.6) \qquad a_{k_j} = 0, \qquad j = 1, \ldots, s.$$

Then the function

$$h(z) = \operatorname{Re} \sum_{\substack{k=0 \\ k \neq k_j}}^{\infty} \frac{a_k}{S_n(R^k)} z^k$$

is harmonic in E_R and the general solution of (D) for E_R and the boundary values f is given by

$$w = S_n h + \sum_{j=1}^{s} \operatorname{Re} \alpha_j S_n(z^{k_j}), \qquad \alpha_j \in \mathbb{C}.$$

(D) is not solvable whenever (2.6) is not fulfilled.

We note that the Dirichlet problem for circles with different centres can be reduced to Theorem 2.1. This follows from the invariance of (0.1) w.r.t. suitable (conformal) transformations. This remark applies in particular to the exterior of circles since $\eta = 1/z$ leaves (0.1) invariant ($\varepsilon = 1$).
R. Heersink studied the linkage of functions in $\Omega_n'(E_R)$ and $\Omega_n'(\mathbb{C}\setminus\overline{E_R})$. Because of the above mentioned properties the results are the expected ones.

2.2. For equation (1.9) problem (D) for circles E_R ($R < 1$ if $\varepsilon = -1$ and $R < t_n$ for $\varepsilon = 1$) can be reduced to the case treated in 2.1. The limiting case $R = 1$ is of special interest, however. It is very remarkable that (D) is solvable also for that circle which is a singular line of (1.9). In fact, the solution is particularly simple and a nice extension of Poisson's formula (cf. [55]): if the in E harmonic function h has the prescribed values on $|z| = 1$ then $w = F_n h$ is the uniquely determined solution of (D) for the same boundary values and E. It has the following explicit representation

$$F_n h = \frac{1}{2\pi S_n(1)} \int_0^{2\pi} f(e^{i\varphi}) \left[\frac{1-|z|^2}{|e^{i\varphi}-z|^2} \right]^{n+1} d\varphi \, . \tag{2.7}$$

Beside being a generalization of Poisson's formula ($n = 0$!) (2.7) extends the second Laplace integral for Legendre functions (the case $f \equiv 1$, cf.[67], p. 314). (2.7) is based on the relation

$$(2.8)\qquad \mathrm{Re}\, S_n\Big(\frac{e^{i\varphi}+z}{e^{i\varphi}-z}\Big) = \Big(\frac{1-|z|^2}{|e^{i\varphi}-z|^2}\Big)^{n+1}$$

and Poisson's theorem for harmonic functions.
With a different kernel (2.7) becomes an analogon to the Schwarz formula of function theory: let $w \in \Omega_n'(E)$, Re w continuous $\bar{E}$. Then

$$w(z) = \frac{1}{2\pi S_n(1)} \int_0^{2\pi} \mathrm{Re}\, w(e^{i\varphi}) S_n\Big(\frac{e^{i\varphi}+z}{e^{i\varphi}-z}\Big)\, d\varphi + i\, \mathrm{Im}\, w(0)$$

for $z \in E$. For further generalizations of the Schwarz formula for (0.1) cf. Bauer [5].

2.3. In this connexion we should mention a valuable mean value property of the functions in $\Omega_n(G)$ (see [59]): let $w \in \Omega_n(G)$, $z_o \in G$, and t' the distance from z_o to ∂G (in the metric of the respective geometry). Then for $r < t'$ $(\varepsilon = -1)$ or $r < \min\{t', t_n\}$ $(\varepsilon = 1)$ we have

$$(2.9)\qquad P_n\Big(\frac{1-\varepsilon r^2}{1+\varepsilon r^2}\Big)\, w(z_o) = \frac{1}{2\pi}\int_0^{2\pi} w\Big(\frac{z_o + re^{i\varphi}}{1-\varepsilon \bar{z}_o\, re^{i\varphi}}\Big)\, d\varphi\ .$$

For $\varepsilon = 1$ (2.9) holds also for $r = t_n$ if the corresponding circle is contained in G. For every such circle we have

$$(2.10)\qquad \int_0^{2\pi} w\Big(\frac{z_o + t_n e^{i\varphi}}{1-\bar{z}_o\, t_n e^{i\varphi}}\Big)\, d\varphi = 0.$$

It should be noted that (2.9) is also sufficient for a twice continuously differentiable function w to be in $\Omega_n(G)$.

CHAPTER 3

Functions with Restricted Range, Schwarz Lemma

3.1. The most important instrument for the representation of positive harmonic functions in E is Herglotz' formula: for every such function $h(z)$ with $h(0) = 1$ there exists a uniquely determined probability measure μ on $[0,2\pi]$ such that

$$h(z) = \int_0^{2\pi} \frac{1-|z|^2}{|e^{i\varphi}-z|^2}\, d\mu(\varphi) , \qquad z \in E.$$

This formula immediately extends to the solutions of (0.1) for $\varepsilon = -1$. In fact, an application of (2.8) shows

$$(S_n h)(z) = \int_0^{2\pi} \left(\frac{1-|z|^2}{|e^{i\varphi}-z|^2}\right)^{n+1} d\mu(\varphi). \tag{3.1}$$

On the other hand, if $w = \mathrm{Re}\, S_n g \in \Omega_n(E)$ is positive in E and fulfils $w(0) = 1$ then Theorem 1.4 gives $h = \mathrm{Re}\, g > 0$ in E.

<u>THEOREM 3.1.</u> ($\varepsilon = -1$). The operator

$$\mu \to \int_0^{2\pi} \left(\frac{1-|z|^2}{|e^{i\varphi}-z|^2}\right)^{n+1} d\mu(\varphi) \tag{3.2}$$

generates a one-to-one mapping between the probability measures on $[0,2\pi]$ and the positive solutions $w \in \Omega_n(E)$ with $w(0) = 1$.

<u>REMARK.</u> Let $w \in \Omega_n(E)$ be positive. Then

$$w(0)\left(\frac{1-|z|}{1+|z|}\right)^{n+1} \leq w(z) \leq w(0)\left(\frac{1+|z|}{1-|z|}\right)^{n+1} \tag{3.3}$$

for $z \in E$ with equality only for the functions

$$w(z) = \alpha\left(\frac{1-|z|^2}{|\eta-z|^2}\right)^{n+1} , \qquad \alpha > 0, \quad |\eta| = 1.$$

For $n = 0$ (3.3) reduces to a well known and important estimate for

positive harmonic functions.
For positive solutions in E_R, $R < t_n'$ [1], we have similar results. This follows from the relation

$$w_n(z,R) = \text{Re } S_n(1 + 2 \sum_{k=1}^{\infty} \frac{S_n(R^0)}{S_n(R^k)} z^k) > 0, \quad z \in E_R,$$

which is a consequence of the solvability of Dirichlet's problem and the maximum principle valid for $\Lambda_n(E_R)$. These considerations are similar to those in the proof of Theorem 1.4.

THEOREM 3.2. The operator

$$\mu \to \int_0^{2\pi} w_n(ze^{-i\varphi},R)d\mu(\varphi)$$

generates a one-to-one mapping between the probability measures on $[0,2\pi]$ and the positive solutions $w \in \Omega_n(E_R)$, $R < t_n'$, with $w(0) = 1$.

In the case $\varepsilon = 1$ an interesting problem arises in connection with positive solutions: it follows from (2.10) that there is no domain of the Riemann sphere which contains a disc of (spherical) radius t_n and admits a positive solution of (0.1). This leads to the question to characterize maximal domains which carry positive solutions. One may conjecture that the largest (w.r.t. to spherical area) spherically convex domain of this type is the spherical disc with radius t_n. This problem remains open.
We should mention that T. Rüdiger used Theorem 3.1 (and its extensions, see [56]) to solve a problem of A. Huber [31] dealing with the representation of positive solutions of the GASP equation (0.4) in the upper halfplane.

3.2. Now we turn to bounded solutions in $\Omega_n'(E_R)$ where similar methods can be applied.

---

1) $$t_n' = \begin{cases} 1, & \varepsilon = -1 \\ t_n, & \varepsilon = 1 \end{cases}$$

<u>THEOREM 3.3.</u> Let $R < t_n'$. For $w \in \Omega_n'(E_R)$ we have [1]

$$\overline{\lim_{r \to R-0}} \; M(r,w) \leq 1$$

if and only if there exists an analytic g in E with $|g(z)| \leq 1$ in E such that

$$w = S_n\left(g * \sum_{k=0}^{\infty} \frac{z^k}{S_n(R^k)}\right) . \tag{3.4}$$

In this case we have

$$M(r,w) \leq S_n(r^o)/S_n(R^o), \qquad 0 \leq r < R.$$

In the limiting case $\varepsilon = -1$, $R = 1$ similar results hold even for the larger class $\Omega_n''(E)$.

<u>THEOREM 3.4.</u> ($\varepsilon = -1$). Let g be analytic in E. Then $|E_n g| \leq E_n(z^n)$ holds in E if and only if $|g(z)| \leq 1$ in E.

To prove Theorem 3.4 we use an integral version of (1.13): Let g be analytic in E, continuous in $\bar{E}$ and assume $|g(z)| \leq 1$, $z \in E$. Then

$$|E_n g| = \left| \frac{n!}{2\pi} \int_0^{2\pi} g\left(\frac{e^{i\varphi}+z}{1+\bar{z}e^{i\varphi}}\right) \frac{(1+\bar{z}e^{i\varphi})^{2n}}{(1-|z|^2)^n} e^{-in\varphi} d\varphi \right| \leq \frac{n!}{2\pi} \int_0^{2\pi} \left(\frac{|1+\bar{z}e^{i\varphi}|^2}{1-|z|^2}\right)^n d\varphi ,$$

with equality if and only if $g(z) = e^{i\gamma} z^n$, $\gamma \in \mathbb{R}$. If g is not continuous on $\bar{E}$ a limiting process can be used.
Now let $|E_n g| \leq E_n(z^n)$. The functions

$$g(z,R) = z^n(1-R^2)^n \sum_{k=0}^{n} \frac{(2n-k)!}{k!(n-k)!} \left(\frac{R^2}{1-R^2}\right)^{n-k} z^{k-n} g^{(k)}(z)$$

satisfy

---

[1] $M(r,w) = \max_{|z|=r} |w(z)|$.

$$\lim_{R \to 1-0} g(z,R) = \frac{(2n)!}{n!} g(z),$$

$$|g(z,R)| \leq R^n(1-R^2)^n E_n R^n, \quad z \in E_R ,$$

such that

$$\frac{(2n)!}{n!} |g(z)| \leq \lim_{R \to 1-0} R^n(1-R^2)^n E_n R^n = \frac{(2n)!}{n!} .$$

Hence $|g(z) \leq 1$ in E.

3.3. One of the most striking properties of analytic functions is the validity of Schwarz' Lemma: For g analytic with $|g(z)| \leq 1$ in E_R and $g(0) = g'(0) = \ldots = g^{(m-1)}(0) = 0$ the estimate

$$|g(z)| \leq |z/R|^m , \quad z \in E_R ,$$

holds. It is very common (and is suggested by the usual method of proof) to consider this Lemma as a consequence of divisibility properties of the analytic functions which, in general, are not shared by solutions of other equations. For $\Omega_n'(E_R)$ we conjecture the existence of similar results and some progress in this direction will be discussed. Furthermore, our methods suggest the existence of such theorems for many other equations. Because of technical difficulties complete information is not yet available.
The case $\varepsilon = -1$, $R = 1$ is once more seperate from the others: the analogon of Schwarz Lemma can be settled by the means of a special method which is completely different from our ideas concerning the general case.

<u>THEOREM 3.5.</u> ($\varepsilon = -1$). Let $w \in \Lambda_n'(E)$ such that

i) $|w(z)| \leq 1, \quad z \in E,$

ii) $\frac{\partial^j w}{\partial z^j}(0) = 0, \quad j = 0, \ldots, m-1.$

Then

(3.5) $$|w(z)| \leq |F_n(z^m)|, \quad z \in E.$$

Note that (3.5) is sufficient for the properties i), ii) of the Theorem.

Proof. Let $w = F_n g$, g analytic in E. Condition ii) gives $g^{(j)}(0) = 0$, $j = 0, \ldots, m-1$, and from Theorem 3.4 we deduce $|g(z)| \leq 1$, $z \in E$. Using $f = z^{-m}g$ we obtain from that Theorem

$$|S_k f| \leq S_k(1), \quad k = 0, \ldots, n, \quad z \in E,$$

and (1.17) gives

$$|z^{-m}S_n g| \leq |S_n f| + \sum_{k=0}^{n-1} \frac{m}{n-k} |S_k f||z^{-m}S_{n-k-1}(z^m)|$$

$$\leq S_n(1) + \sum_{k=0}^{n-1} \frac{m}{n-k} S_k(1) z^{-m} S_{n-k-1}(z^m)$$

$$= |z^{-m}S_n(z^m)|.$$

The estimate follows after multiplication with $|z^m/S_n(1)|$.

<u>CONJECTURE A.</u> Let $R < t_n'$. For $w \in \Omega_n'(E_R)$ assume

$$\text{i)} \quad \overline{\lim_{r \to R-0}} \; M(r,w) \leq 1,$$

$$\text{ii)} \quad \frac{\partial^j w}{\partial z^j}(0) = 0, \quad j = 0, \ldots, m-1.$$

Then

(3.6) $$M(r,w) \leq S_n(r^m)/S_n(R^m), \quad 0 \leq r < R.$$

We shall prove this Conjecture for $n = 1$. A necessary and sufficient condition for its truth (for $n \in \mathbb{N}$) is given in the next Theorem.

THEOREM 3.6. For n, m, R fixed Conjecture A holds if and only if

$$\operatorname{Re} b_r(z) > 1/2, \quad z \in E, \tag{3.7}$$

where

$$b_r(z) = \sum_{k=0}^{\infty} \frac{s_n(R^m)}{s_n(r^m)} \frac{s_n(r^{m+k})}{s_n(R^{m+k})} z^k, \quad 0 < r < R.$$

Proof. Conditions i), ii) of Conjecture A are fulfilled if and only if

$$w(z) = \sum_{k=m}^{\infty} b_{k-m} \frac{s_n(z^k)}{s_n(R^k)}, \quad z \in E_R,$$

where

$$g(z) = \sum_{k=0}^{\infty} b_k z^k \in \Omega_o'(E)$$

satisfies $|g(z)| \leq 1$, $z \in E$ (Theorem 3.3). The Schwarz Lemma implies (3.6) if and only if $|(g * b_r)(z)| \leq 1$, $z \in E$, $0 < r < R$, for all admissable g. But this is equivalent to (3.7) (cf. T. Sheil-Small [63]).

THEOREM 3.7. Conjecture A holds for n = 1.

Proof. We have to show that for $0 < r < R$, $\varphi \in \mathbb{R}$,

$$G(m) = \sum_{k=0}^{\infty} \frac{k+m+d(r)}{k+m+d(R)} \left(\frac{r}{R}\right)^k \cos(k\varphi) - \frac{1}{2} \frac{m+d(r)}{m+d(R)} > 0,$$

where

$$d(x) = (1-\varepsilon x^2)/(1+\varepsilon x^2).$$

It follows from Theorems 3.3, 3.6 that $G(0) > 0$. Furthermore, $\lim_{m\to\infty} G(m) > 0$. Hence it suffices to show that $G'(m)$ has constant sign. We have

$$G'(m) = (d(R) - d(r))(m+d(R))^{-2} H(m)$$

with

$$H(m) = \sum_{k=0}^{\infty} \left(\frac{m+d(R)}{k+m+d(R)}\right)^2 \left(\frac{r}{R}\right)^k \cos(k\varphi) - \frac{1}{2}.$$

The coefficients of H(m) form a positive, decreasing and convex sequence and by the above mentioned result of Rogosinski we conclude: $H(m) > 0$.
Now let $\varepsilon = 1$. Our investigations in the proof of (2.2) show that (3.7) holds for all $R < \infty$, $n \in \mathbb{N}$, and large $m \in \mathbb{N}$. Therefore, Theorem 3.8 follows from Theorem 3.6.

THEOREM 3.8. For every $R < \infty$ there exists $m_o(R,n)$ such that for $m \geq m_o$ and $w \in \Omega_n'(E_R)$ with

$$\text{i)} \quad \overline{\lim_{r \to R-0}} \; M(r,w) \leq 1,$$

$$\text{ii)} \quad \frac{\partial^j w}{\partial z^j}(0) = 0, \quad j = 0, \ldots, m-1,$$

we have

$$M(r,w) \leq S_n(r^m)/S_n(R^m), \quad 0 \leq r < R.$$

For $n = 1$ we can show $m_o(R,1) = 1$ for $R \geq t_1$. This gives reason to the more general

CONJECTURE B. ($\varepsilon = 1$).

$$m_o(R,n) = \begin{cases} 0 \text{ for } R < t_n, \\ n \text{ for } R \geq t_n. \end{cases}$$

A considerable amount of numerical experiments supports Conjecture B. Theorem 3.8 contains an interesting conclusion: under the same assumptions the function $M(r,w)/S_n(r^m)$ is monotonically increasing for $0 < r < R$ and $m \geq m_1(R,n) = \max\{m_o(R,n) \mid r < R\}$. On the other hand, it can be shown that for $m > n$:

$$\frac{d}{dr} S_n(r^m) = \left(\frac{r^2}{1+r^2}\right)^n \sum_{j=-1}^{n} b_j r^{m-2j-1}$$

with nonnegative coefficients b_j. Thus $S_n(r^m)$ itself is monotonically increasing in r if $m > n$. Hence we have:

THEOREM 3.9. ($\varepsilon = 1$). Let $w \in \Omega_n'(E_R)$ and

$$\frac{\partial^j w}{\partial z^j}(0) = 0, \quad j = 0, \ldots, m-1,$$

for $m > m_2(R,n) = \max\{m_1(R,n), n\}$. Then $M(r,w)$ is monotonically increasing for $0 < r < R$.

In case that Conjecture B holds we can choose $m_2(R,n) = n$ independently of R (and this would be best possible as the example $S_n(z^n)$ shows). This would enable us to draw interesting conclusions, in particular for 'entire solutions'. It should be noted that for $\varepsilon = 1$ there is no general maximum principle.

CHAPTER 4

Univalent Solutions, Riemann Mapping Theorem

4.1. In this chapter we shall deal with univalent solutions in $\Omega_n'(E)$ and $\Lambda_n'(E)$ for $\varepsilon = 1$. Particularly this part shows the power of function theoretic methods if applied to related problems for certain partial differential equations. We shall need some notations and facts from the 'geometric' theory of functions.
An analytic function $f(z)$ with $f'(0) = 0$ maps E univalently onto a convex domain if and only if

$$(4.1) \qquad \operatorname{Re} \frac{zf''(z)}{f'(z)} > -1, \qquad z \in E.$$

If $f(0) = 0$ it maps E univalently onto a domain 'starlike w.r.t. the origin' if and only if

$$(4.2) \qquad \operatorname{Re} \frac{zf'(z)}{f(z)} > 0, \qquad z \in E.$$

These functions are called 'convex' or 'starlike' in E and corresponding results hold for other circles E_r. The following result is in [61]:

THEOREM 4.A. Let f be convex, g starlike in E. Then, if P is analytic in E with $P(0) = 1$ and $\operatorname{Re} P > 0$ in E we have

$$(4.3) \qquad \operatorname{Re} \frac{f*(gP)}{f*g} > 0, \qquad z \in E.$$

In particular, $f * g$ is starlike in E.

Since f convex if and only if zf' starlike we conclude from Theorem 4.A: $f*g$ convex if f, g convex (Pólya-Schoenberg conjecture).
Let f, g be analytic in E, $f(0) = g(0)$. Then f is called 'subordinate' to g ($f \prec g$) if there exists an analytic function u in E with $u(0) = 0$ and $|u(z)| < 1$ such that $f(z) = g(u(z))$ in E. For g convex we have $f \prec g$ if and only if

$$(4.4) \qquad \operatorname{Re} \frac{zg'(z)}{g(z)-f(\sigma)} > 0, \qquad |\sigma| < |z| < 1.$$

Together with Theorem 4.A this leads to

$$(4.5) \qquad f \prec g \Rightarrow f * h \prec g * h$$

whenever g, h are convex.
A function $f(z,t)$, $(z,t) \in E \times [a,b]$, is called 'subordination chain' on $[a,b]$ if for all $a < s < t < b$ the following properties are fulfilled [1)]:

i) $f(z,t)$ analytic in E,
ii) $f(0,t) = 0, \quad f'(0,t) > 0$,
iii) $f(z,s) \prec f(z,t)$.

<u>THEOREM 4.B.</u> (Pommerenke [45]). Let $f(z,t)$ be a subordination chain on $[a,b]$, locally absolutely continuous, locally uniformly in E. Then the relation

$$(4.6) \qquad \operatorname{Re} \frac{\dot{f}(z,t)}{zf'(z,t)} > 0, \quad z \in E,$$

holds for almost all $t \in [a,b]$. On the other hand, if $f(z,t)$ is univalent and continuous in $\bar{E}$ for all $t \in [a,b]$ and fulfils

$$(4.7) \qquad \operatorname{Re} \frac{\dot{f}(z,t)}{zf'(z,t)} > \delta > 0, \quad z \in E, \quad t \in [a,b],$$

then $f(\bar{E},s) \subset f(E,t)$ for all $a < s < t < b$.

Fundamental for the considerations of this chapter is the following property of the functions $F_n(1/(1-z))$. The lengthy proof is in [60]. For the notations see (1.22).

<u>THEOREM 4 1.</u> ($\varepsilon = -1$). For every $t \in (0,1)$ there exists $r_o = r_o(t,n) > 1$ such that

$$f_n(z,t;z/(1-z)) = \sum_{k=1}^{\infty} F_n(t^k) z^k$$

is convex in $|z| < r_o$.

---

1) $f'(z,t) = \frac{\partial}{\partial z} f(z,t), \quad \dot{f}(z,t) = \frac{\partial}{\partial t} f(z,t)$

Now let g be convex in E. Theorems 4.A and 4.1 show that $f_n(z,t;g)$ is convex in E_{r_o} and from Theorem 1.3 applied to $w = F_n g$ and $s < t$ we deduce

$$w(\partial E_s) \subset w(E_t) \subset f_n(E,t;g).$$

Since $w(\partial E_t) = f_n(\partial E,t;g)$ we obtain $w(\partial E_s) \cap w(\partial E_t) = \emptyset$. Together with the univalence of $f_n(z,t;g)$ on ∂E this shows that $w = F_n g$ is univalent in E. An application of Caratheodory's kernel Theorem gives $w(E) = g(E)$.

Let $w = F_n g$, g analytic in E, be univalent in E. Without loss of generality we assume $w(0) = g(0) = 0$, $w_z(0) = (n+1)g'(0) > 0$. Obviously $f_n(z,t;g)$ forms a subordination chain on $t \in (0,1)$ which satisfies the assumptions of Theorem 4.B in every closed subinterval. Therefore

$$\text{(4.8)} \qquad \operatorname{Re} \frac{\dot{f}_n(z,t;g)}{z f_n'(z,t;g)} \geq 0, \quad z \in E, \quad 0 < t < 1,$$

and from (1.26), for $t \to 1-0$, we obtain

$$\dot{f}_n(z,t;g) = \frac{1-t}{2n+1} z(zg')' + O((1-t)^2),$$

$$z f_n'(z,t;g) = zg' + O((1-t)^2).$$

Thus the limit $t \to 1-0$ in (4.8) gives

$$\operatorname{Re} \frac{zg''(z)}{g'(z)} > -1, \quad z \in E,$$

and we conclude: g convex in E.

THEOREM 4.2. ($\varepsilon = -1$). i) Let $w \in \Lambda_n'(E)$ be univalent. Then $w(E)$ is a convex domain.
ii) Let G be a convex domain in $\mathbb{C}$, $G \neq \mathbb{C}$, and let $w_o \in G$. Then there exists a uniquely determined univalent function $w \in \Lambda_n'(E)$ with $w(E) = G$, $w(0) = w_o$ and $w_z(0) > 0$.

This Theorem 4.2 is similar to a version of the Riemann Mapping Theorem. Note that only convex domains appear as images of univalent mappings in $\Lambda_n'(E)$.

From the proof of Theorem 4.2 two other conclusions can be obtained.

1) Let $v = F_n g$, $w = F_n h$ be univalent functions in $\Lambda_n'(E)$. Then the function

$$u = v*w: = F_n(g*h)$$

has the same property.

2) If $w \in \Lambda_n'(E)$ is univalent in E then the images $w(E_t)$, $0 < t < 1$, are convex domains.

4.2. In this section we study the Jacobian

$$|w_z|^2 - |w_{\bar{z}}|^2$$

for univalent functions in $\Lambda_n'(E)$. If $zw_z \neq \bar{z}w_{\bar{z}}$ we have for $w = F_n g$

$$\text{(4.9)} \qquad \frac{zw_z + \bar{z}w_{\bar{z}}}{zw_z - \bar{z}w_{\bar{z}}} = z\,\frac{\dot{f}_n(e^{i\varphi},|z|;g)}{e^{i\varphi}f_n'(e^{i\varphi},|z|;g)}, \qquad z = |z|e^{i\varphi}.$$

Therefore the condition $|w_z| > |w_{\bar{z}}|$ implies

$$\text{(4.10)} \qquad \mathrm{Re}\,\frac{\dot{f}_n(e^{i\varphi},|z|;g)}{e^{i\varphi}f_n'(e^{i\varphi},|z|;g)} > 0, \qquad z \in E,$$

and, in particular, $f_n'(\eta,z;g) \neq 0$, $|\eta| = 1$. By means of the argument principle we can see that this is true also for $\eta \in E$ otherwise there would be a contradiction to $w_z - (\bar{z}/z)w_{\bar{z}} \neq 0$. Thus (4.10) implies (4.8) and we obtain: g convex in E.

THEOREM 4.3. ($\varepsilon = -1$). Let $|w_z| > |w_{\bar{z}}|$, $z \in E$, for $w \in \Lambda_n'(E)$, $n \in \mathbb{N}$. Then w is univalent in E.

Note that this Theorem is false for $n = 0$: an analytic function which is locally univalent in E is not necessarily univalent in E.

One may conjecture that every univalent function in $\Lambda_n'(E)$, $n \in \mathbb{N}_0$, has nonvanishing Jacobian in E, i.e. it is a diffeomorphism. For $n = 1$ this is true:

THEOREM 4.4. ($\varepsilon = -1$). Let $w \in \Lambda_1'(E)$ be univalent in E. Then

$$(4.11) \qquad |w_{\bar z}| \le \frac{1+|z|}{1+|z|^5}\,|z|^2|w_z| < |w_z| .$$

Proof. Because of (4.9) and Theorem 4.A it is sufficient to verify this estimate for $w = F_n(z/(1-z))$. For $n = 1$ this gives (4.11).

4.3. We want to establish a number of applications of our previous results.
1) The following Theorem extends a well known subordination for analytic functions.

THEOREM 4.5. ($\varepsilon = -1$). Let v, $w \in \Lambda_n'(E)$, w univalent. Assume

$$(4.12) \qquad v(0) = w(0), \quad v(E) \subset w(E).$$

Then $v(E_t) \subset w(E_t)$ holds for all $t \in (0,1)$.

Proof. Let $v = F_n H$, $w = F_n g$. From Theorems 1.4 and 4.2 we obtain $h \prec g$, g convex. (4.5) gives

$$f_n(z,t;h) \prec f_n(z,t;g), \quad |z| < r_0(t,n),$$

which implies

$$f_n(E,t;h) \subset f_n(E,t;g).$$

Now Theorem 1.3 shows

$$v(E_t) \subset \dot{\mathrm{clco}}(f_n(E,t;h)) \subset \dot{\mathrm{clco}}(f_n(E,t;g)) = f_n(E,t;g) \subset w(E_t).$$

If we put $w_0 = F_n((1+z)/(1-z))$ and observe $w_0(E) = \{z \mid \mathrm{Re}\, z > 0\}$ then Theorem 4.5 yields the following application: let $v \in \Lambda_n'(E)$, $v(0) = 1$, $\mathrm{Re}\, v(z) > 0$ in E. Then for $0 < t < 1$,

$$v(E_t) \subset w_0(E_t).$$

This generalizes the important subordination $g \prec (1+z)/(1-z)$ for g analytic with $g(0) = 1$, $\mathrm{Re}\, g(z) > 0$ in E.

2) Let $w = F_n g \in \Lambda_n'(E)$ satisfy $|w(z)| < M$ in E and $w_z(0) = 1$. Then $|g(z)| < M$ in E, $g'(0) = 1/(n+1)$, such that there exists a radius $r' = r'(M,n)$ with the property that g is convex in $|z| < r'$.

THEOREM 4.6. ($\varepsilon = -1$). Let $w \in \Lambda_n'(E)$ with $|w(z)| < M$ in E, $w_z(0) = 1$. Then w is univalent in $|z| < r'$. A (rough) estimate for r' is given by

$$r'(M,n) \geq (2-\sqrt{3})[M(n+1) - \sqrt{M^2(n+1)^2-1}].$$

3) From (1.24) we obtain for $t_1, t_2 \to 1-0$, $n \in \mathbb{N}$,

$$\frac{F_n(t_1^k) - F_n(t_2^k)}{F_n(t_1) - F_n(t_2)} \to k^2, \quad k \in \mathbb{N}. \tag{4.13}$$

THEOREM 4.7. ($\varepsilon = -1$). For $z_1, z_2 \in E$ we have

$$\left|\frac{F_n(z_1^k) - F_n(z_2^k)}{F_n(z_1) - F_n(z_2)}\right| \leq k^2, \quad k \in \mathbb{N}. \tag{4.14}$$

Proof. Since $z + (\alpha/k^2)z^k$, $|\alpha| \leq 1$, is convex in E we deduce the univalence of

$$w(z) = F_n(z) + (\alpha/k^2)F_n(z^k), \qquad |\alpha| \leq 1.$$

The result follows from $w(z_1) \neq w(z_2)$, $z_1, z_2 \in E$.

Note that the constant k^2 can be replaced by k for $n = 0$. In general, however, k^2 is best possible as shown by (4.13).

4.4. For the univalent functions in $\Omega_n'(E)$, $\varepsilon = -1$, we have not been able to obtain complete results. Theorem 4.1 proves the convexity of

$$s_n(z,t;z/(1-z)) = \sum_{k=1}^{\infty} s_n(t^k)z^k, \qquad 0 < t < 1,$$

in $|z| < r_o$.

THEOREM 4.8. ($\varepsilon = -1$). Let $w = S_n g \in \Omega_n'(E)$, g convex in E, $g(0) = 0$. Then w is univalent in E.

On the other hand, if $w = S_n g$ is univalent in E with $w(0) = 0$ such that

$$\operatorname{Re} \frac{\mathring{s}_n(z,t;g)}{z s_n'(z,t;g)} \geq 0, \quad z \in E, \quad 0 < t < 1,$$

then the limit $t \to 1-0$ gives

$$\operatorname{Re} \frac{zg'(z)}{g(z)} > 0, \quad z \in E,$$

which shows that g is starlike in E. In fact, there exist univalent functions in $\Omega_n'(E)$ which vanish at $z = 0$ but are not generated by a convex analytic function. Direct calculations with $w = S_1(z+\alpha z^2)$ show that w is univalent at least for $|\alpha| < .4$ but not univalent when $|\alpha| > .48$. On the other hand, $g(z) = z + \alpha z^2$ is convex if and only if $|\alpha| < .25$ and starlike if and only if $|\alpha| < .5$. This proves that the class of univalence generating analytic functions lies between the convex and the starlike class. A precise (in particular geometrical) description of this class appears to be difficult. An analogous result to Theorem 4.3 is true, however.

THEOREM 4.9. ($\varepsilon = -1$). For $w \in \Omega_n'(E)$, $n \in \mathbb{N}$, assume $w(0) = 0$, $|w_z| > |w_{\bar{z}}|$ in E. Then w is univalent in E.

Proof. For such a function $w = S_n g$ we have

$$\operatorname{Re} \frac{\mathring{s}_n(\sigma,|z|;g)}{\sigma s_n'(\sigma,|z|;g)} > 0, \quad \sigma \in E, \quad z \in E.$$

Again, g is starlike and by Theorem 4.A we see that $s_n(z,t;g)$ is starlike in $|z| < r_o$. In particular it is univalent on ∂E. An application of (4.7) for subintervals of $(0,1)$ completes the proof.

Many questions remain open; we mention a few of them.

1) Are univalent solutions $w = S_n g$ in E or E_R diffeomorphisms?

2) Are the sets of univalent solutions $w \in \Omega_n'(E)$, $w(0) = 0$, compact sets?

3) Is it possible to remove the condition $g(0) = 0$ from the assump-

tions in Theorem 4.8?

4) What about the corresponding questions for the linear invariant class $\Omega_n''(E)$?

5) Let $w = \sum_{k=0}^{\infty} a_k S_n(z^k)$ be univalent in E. What are the coefficient estimates, in particular if $a_0 \neq 0$?

4.5. In several papers (see MacGregor [41], Brickman [15], Pólya and Schoenberg [44]) conditions have been obtained which can be described as 'convexity through subordination'. In [44], for instance, it is shown that g is convex in E if and only if for the de la Vallée Poussin means $V_m(z,g)$ the relations

$$V_m(z,g) \prec g, \quad m \in \mathbb{N},$$

hold. The same conclusion is true for the condition [41]

$$\frac{1}{t}\int_0^t g(ze^{i\varphi})d\varphi \prec g, \quad 0 < t < \delta.$$

Our Theorem 4.2 contains a similar result for every $n \in \mathbb{N}$. In the case $n = 1$ we have

$$\frac{1-t^2}{1+t^2}\, tz\, g'(tz) + g(tz) \prec g(z), \quad 0 < t < 1,$$

if and only if g convex in E. The function on the left hand side is also convex and forms a subordination chain.

Another related problem comes from (1.15): let g (and therefore $f_n(z,t;g)$) be convex in E. Then (1.15) and (4.8) give

$$\operatorname{Re} \frac{zf_n'(z,t;g)}{f_{n+1}(z,t;g)-f_n(z,t;g)} > 0, \quad z \in E, \quad 0 < t < 1.$$

Comparing this with (4.4) the following conjecture is suggested:

$$f_n(z,t;g) \prec f_{n+1}(z,t;g), \quad n \in \mathbb{N}.$$

Again, it would be sufficient to decide this for $g = z/(1-z)$.

CHAPTER 5

Spaces of Hardy Type

5.1. This chapter is devoted to explain a few basic facts on Hardy Spaces for the solutions of (1.9), $\varepsilon = -1$. The complete proofs are in [56].

The solvability of Dirichlet's problem and (1.18) show that

$$(5.1) \qquad M_p(r,w) = \left(\frac{1}{2\pi}\int_0^{2\pi} |w(re^{i\varphi})|^p d\varphi\right)^{1/p}, \qquad p \geq 1,$$

increases monotonically in $0 < r < 1$ if $w \in \Lambda_n(E)$. Similar to the classical case we define for $0 < p < \infty$: $w \in \Lambda_n(E)$ is in h_n^p if and only if

$$\|w\|_p = \sup_{0<r<1} M_p(r,w) < \infty .$$

For $p \geq 1$ we have

$$\|w\|_p = \lim_{r\to 1-0} M_p(r,w).$$

For a harmonic function h in E, $n \in \mathbb{N}_0$ let

$$w_n(z) = F_n h = \sum_{k=-\infty}^{\infty} a_k F_n(r^{|k|})e^{ik\varphi}, \quad z = re^{i\varphi},$$

such that for $t \in [0,1]$ and $m \in \mathbb{N}_0$

$$w_{nt}(z) = \sum_{k=-\infty}^{\infty} a_k F_m(t^{|k|})F_n(r^{|k|})e^{ik\varphi} \in \Lambda_n(E).$$

For $p \geq 1$ we obtain

$$M_p(r,w_{nt}) \leq M_p(1,w_{nt}) = M_p(t,w_m)$$

and thus

$$M_p(r,w_n) \leq \|w_m\|_p .$$

Interchanging of m, n completes the proof of the following Theorem.

THEOREM 5.1. ($\varepsilon = -1$). Let $p \geq 1$. $w = F_n h$ is in h_n^p if and only if $h \in h_o^p$ and in this case $\|F_n h\|_p = \|h\|_p$.

As an immediate consequence of Theorem 5.1 we see that h_n^p, $p \geq 1$, are Banach spaces w.r.t. their norms.
From the Poisson-Stieltjes representation for h_o^p, Theorem 5.1 and (2.8) we conclude:

THEOREM 5.2. ($\varepsilon = -1$). For every $w \in h_n^p$, $p \geq 1$, there is a $\hat{w} \in L^p[0,2\pi]$ $(p > 1)$ or a finite Baire measure μ on $[0,2\pi]$ such that

$$w(z) = \frac{1}{2\pi}\int_0^{2\pi}\left(\frac{1-|z|^2}{|e^{i\varphi}-z|^2}\right)^{n+1}\hat{w}(e^{i\varphi})d\varphi, \qquad p > 1, \tag{5.2}$$

or

$$w(z) = \frac{1}{2\pi}\int_0^{2\pi}\left(\frac{1-|z|^2}{|e^{i\varphi}-z|^2}\right)^{n+1}d\mu(\varphi), \qquad p = 1.$$

An application of the Bohman-Korovkin Theorem (cf.[16], Th. 1.3.7) to the positive kernel of these integrals results in

$$\lim_{r\to 1-0}\int_0^{2\pi}|w(re^{i\varphi}) - \hat{w}(e^{i\varphi})|^p d\varphi = 0, \qquad p > 1.$$

Fatou's Theorem and Littlewood's Maximality Theorem can also be extended, see [56].
In the same manner we can connect the Hardy spaces $H_n^p = h_n^p \cap \Lambda_n'(E)$ with the spaces $H^p = H_o^p$ of analytic functions in E. This yields results similar to the Theorems of M. Riesz and Kolmogoroff on conjugate harmonic functions. We mention an important special case.

THEOREM 5.3. ($\varepsilon = -1$). Let $w \in \Lambda_n'(E)$, $\mathrm{Re}\, w > 0$. Then $w \in H_n^p$ for $p \in (0,1)$.

5.2. Every univalent analytic function in E belongs to H_o^p for all $p \in (0,1/2)$ (Prawitz [46]). The corresponding result in $\Lambda_n'(E)$ is

THEOREM 5.4. ($\varepsilon = -1$). Let $w \in \Lambda_n'(E)$ be univalent in E. Then $w \in H_n^p$ for all $p \in (0,1)$.

Proof. Let $w = F_n g$. Then g is convex and a theorem of Eenigenburg and Keogh [20] shows $g \in H_o^{1+\delta}$, $\delta > 0$, or $g(z) = (a+bz)/(1+cz)$, $|c| = 1$. In the first case $w \in H_n^1 \subset H_n^p$, $0 < p < 1$, otherwise the result follows from Theorem 5.3.

We also mention a generalization of Rogosinski's subordination Theorem for H_o^p.

THEOREM 5.5. ($\varepsilon = -1$). Let $v, w \in \Lambda_n'(E)$, $v(0) = w(0)$, w univalent in E with $v(E) \subset w(E)$. Then for $1 \leq p < \infty$

$$M_p(r,v) \leq M_p(r,w), \quad 0 < r < 1.$$

In fact, as in Theorem 4.5 we conclude $f_n(z,t;h) \prec f_n(z,t;g)$ for $v = F_n h$, $w = F_n g$. The result follows from Rogosinski's Theorem.

CHAPTER 6

Summability, Abel's Theorem

6.1. A series

$$\sum_{k=0}^{\infty} a_k \tag{6.1}$$

is called summable by Abel's method A to the sum s if

$$\lim_{t\to R-0} \sum_{k=0}^{\infty} a_k \frac{t^k}{R^k} = s$$

for a certain $R > 0$. Let $R < 1$ for $\varepsilon = -1$ and $S_n(R^k) \neq 0$, $k \in \mathbb{N}_0$, for $\varepsilon = 1$. The series (6.1) is called summable by the generalized Abel's method A' to the sum s if

$$\lim_{t\to R-0} \sum_{k=0}^{\infty} a_k \frac{S_n(t^k)}{S_n(R^k)} = s.$$

For $n = 0$ we have $A = A'$. In fact, more is true:

THEOREM 6.1. A series is A' summable if and only if it is A summable. The sums are equal.

Proof. For both, A and A', it is necessary that

$$h(z) = \sum_{k=0}^{\infty} a_k \frac{z^k}{S_n(R^k)}$$

is analytic in E_R. From (1.31) we deduce

$$\sum_{k=0}^{\infty} a_k \frac{S_n(t^k)}{S_n(R^k)} = (S_n h)(t) = \sum_{j=0}^{n} p_j(t) t^j h^{(j)}(t)$$

where $p_j(t)$, $j = 1, \ldots, n$, are analytic in $0 \leq t \leq R$ and $p_n(t) \equiv 1$. On the other hand

$$\sum_{k=0}^{\infty} a_k \frac{t^k}{R^k} = \sum_{j=0}^{n} p_j(R)t^j h^{(j)}(t),$$

and it remains to show:

$$\sum_{j=0}^{n} p_j(t)t^j h^{(j)}(t) = s + o(1), \quad t \to R-0, \tag{6.2}$$

if and only if

$$\sum_{j=0}^{n} p_j(R)t^j h^{(j)}(t) = s + o(1), \quad t \to R-0. \tag{6.3}$$

The existence and uniqueness theorems for linear differential equations can be used to show that both conditions, (6.2) and (6.3), imply

$$h^{(j)}(t) = O(1), \quad j = 0, \ldots, n, \quad t \to R-0.$$

Since $p_j(t) \to p_j(R)$ the result follows.

The next Theorem generalizes Abel's Theorem. It is a Corollary to Theorem 6.1.

<u>THEOREM 6.2.</u> Let the series (6.1) be convergent and R as above. Then we have

$$\lim_{t\to R-0} \sum_{k=0}^{\infty} a_k \frac{S_n(t^k)}{S_n(R^k)} = \sum_{k=0}^{\infty} a_k .$$

It is easily seen that the continuity statement of Theorem 6.2 is extendable into Stolz angles in E_R with vertex in $z = R$.

6.2. We turn to the case $\varepsilon = -1$, $R = 1$. Let $\{a_k\}$ be such that

$$h(z) = \sum_{k=0}^{\infty} a_k z^k$$

is analytic in E. We define a summability method [1)]

$$\sum_{k=0}^{\infty} a_k = s \qquad (A_n)$$

by the condition

$$\lim_{r\to 1-0} \frac{(1-r^2)^n}{\binom{2n}{n}} (S_n h)(r) = s.$$

<u>THEOREM 6.3.</u> ($\varepsilon = -1$). Let h be harmonic in E. Then

$$\lim_{r\to 1-0} \frac{(1-r^2)^n}{\binom{2n}{n}} (S_n h)(r) = s \tag{6.4}$$

if and only if

$$\begin{aligned} &\text{i)} \quad \lim_{r\to 1-0} h(r) = s, \\ &\text{ii)} \quad \lim_{r\to 1-0} (1-r)^n h^{(n)}(r) = 0. \end{aligned} \tag{6.5}$$

Proof. It is enough to consider the case $s = 0$. From (6.4) and (1.14) we obtain

$$(S_{n-1}h)(r) + \frac{r}{n}\,\frac{1-r^2}{1+r^2}\,\frac{\partial}{\partial r}\,(S_{n-1}h)(r) = o((1-r)^{-n+1}). \tag{6.6}$$

Let $y = r/(1-r^2)$ and $(S_{n-1}h)(r) = y^{n-2}g(y)$ such that (6.6) becomes

$$\frac{2n-2}{y}\, g(y) + g'(y) = o(1), \qquad y \to \infty.$$

A theorem of G. Hardy [26] shows $g(y) = o(y)$ or

$$(S_{n-1})(r) = o((1-r)^{-(n-1)}).$$

[1)] Important hints related to the next two theorems have been given by K.H. Indlekofer.

By mathematical induction we can prove

$$(6.7) \qquad (S_k h)(r) = o((1-r)^{-k}), \quad k = 0,1, \ldots, n.$$

Theorem 6.3 holds for n = 0. Assume it holds for all parameters < n. Then (6.4), (6.5) and (6.7) yield

$$(6.8) \qquad \lim_{r \to 1-0} (1-r)^k h^{(k)}(r) = 0, \quad k = 0,1, \ldots, n-1.$$

The necessity of (6.5) for (6.4) follows from (1.31) and (6.4). On the other hand, (6.5) implies (6.8) (see Boas [12]) and thus (6.4). Theorem 6.3 and Boas' result show that the method $A = A_0$ is stronger than A_n ($A \supset A_n$) and, in fact, $A_{n+1} \subset A_n$, $n \in \mathbb{N}_0$. All these methods are consistent.

<u>THEOREM 6.4.</u> ($\varepsilon = -1$). Let

$$\sum_{k=0}^{\infty} a_k = s \; (C_\alpha)$$

for a Cesaro method of order $\alpha > -1$. Then

$$\frac{(1-|z|^2)^n}{\binom{2n}{n}} S_n \left(\sum_{k=0}^{\infty} a_k z^k \right) \to s$$

uniformly in every Stolz angle in E with vertex in z = 1.

Proof. In view of (1.31) it suffices to show that

$$h(z) = \sum_{k=0}^{\infty} a_k z^k$$

fulfils

$$(6.9) \qquad (1-|z|^2)^k h^{(k)}(z) \to \begin{cases} s, & k = 0, \\ 0, & k \in \mathbb{N}, \end{cases}$$

uniformly in the Stolz angles. Since $C_\alpha \subset A$, (6.9), for k = 0, follows from the extended version of Abel's Theorem (Knopp [35]). The cases

$k \in \mathbb{N}$ can be obtained by a careful estimate of Cauchy's integral for small circles in the Stolz angles.

The Theorems above imply: for $\alpha > -1$, $n \in \mathbb{N}$, the relations

$$C_\alpha \subset A_{n+1} \subset A_n \subset A$$

hold with consistency.

6.3. D. Borwein [13] introduced a scale of summability methods A^0_α, $\alpha > -1$, namely

$$s_k \to s \ (A^0_\alpha)$$

if and only if

$$\lim_{t \to 1-0} (1-t)^{\alpha+1} \sum_{k=0}^{\infty} \binom{k+\alpha}{k} s_k t^k = s, \tag{6.10}$$

and proved

$$A^0_\alpha \subset A^0_\beta, \quad \alpha > \beta > -1. \tag{6.11}$$

These methods have been studied in a number of papers, see [14], [43]. Let $\alpha = n$ and

$$s_k = \sum_{j=0}^{k} a_j, \quad h(z) = \sum_{j=0}^{\infty} a_j z^j,$$

such that

$$(1-t)^{n+1} \sum_{k=0}^{\infty} \binom{k+n}{k} s_k t^k = \sum_{m=0}^{n} C_{nm}(t)(1-t)^m h^{(m)}(t), \tag{6.12}$$

where

$$C_{nm}(t) = \sum_{j=m}^{n} \frac{n!\, t^{n-j+m}}{m!\,(m-j)!\,(n+m-j)!} (1-t)^{j-m}$$
$$= \frac{t^n}{m!} + O((1-t)), \quad t \to 1-0.$$

(6.11), (6.12) and Theorem 6.3 show that Borwein's method A_n^o is equivalent to our method A_n.

It is an open question whether Borwein's methods A_α^o are equivalent to the summability methods arising from the equation (0.1) for all $n = \alpha > -1/2$.

6.4. Borwein's methods are power series methods in the sense of Włodarski [68] and therefore perfect, i.e. consistent with every stronger functional (see [68]) summability method. Thus we have:

THEOREM 6.5. ($\varepsilon = -1$). The summability methods A_n, $n \in \mathbb{N}$, are perfect.

In generalization of the power series methods mentioned above one can define summability methods by generalized power series

$$w(z) = \sum_{k=0}^{\infty} a_k S_n(z^k).$$

In fact, if $a_k > 0$, $k \in \mathbb{N}_0$, and $\lim_{t \to R-0} w(t) = \infty$ for a $R < t_n'$ we introduce a permanent summability method for $\{x_k\}$ by

$$\lim_{t \to r-0} \sum_{k=0}^{\infty} \frac{a_k S_n(t^k)}{w(t)} x_k .$$

In extension of Włodarski's work it is possible to show that these methods are perfect. For the proof, which is valid also for more general operators than S_n we refer to [62].

CHAPTER 7

Range Problems

7.1. We shall study the range of functions in $\Omega_n'(E)$, $\varepsilon = -1$. Obviously, beside $w \equiv 0$ there is no bounded function in this class. In fact, even the condition

$$(7.1) \qquad M(r,w) = o((1-r)^{-n}), \quad r \to 1-0,$$

is satisfied only by $w \equiv 0$. On the other hand, $w(E) = \mathbb{C}$ is often the case.

THEOREM 7.1. ($\varepsilon = -1$). Let $w \in \Omega_n'(E)$ be non-constant, $w(0) = 0$. Then $w(E) = \mathbb{C}$.

Proof. Let $w = S_n g$, g analytic with $g(0) = 0$. Then $g(E)$ contains a disc $|\eta| < r(g)$. The functions $f_n(z,t;g)$ converge to $g(z)$, uniformly on compact subsets of E. Thus there is a $t' < 1$ such that $f_n(E,t;g)$ covers the disc $|\eta| < r(g)/2$, $t' < t < 1$. Hence $s_n(E,t;g)$, $t < t' < 1$, covers the disc $|\eta| < r(g)S_n(t^o)/2$, and these discs are contained in $w(E)$. Since $S_n(t^o) \to \infty$ for $t \to 1-0$ the proof is complete.

An assumption of the type $w(0) = 0$ cannot be omitted completely: the function $S_n((1+z)/(1-z))$ takes values only in the right half plane. It is not unlikely, however, that $w(0) = 0$ can be replaced by $0 \in w(E)$.

REMARK. Combining the previous Theorem 7.1 with the results of Chapter 4 gives: for g convex, $g(0) = 0$, the function $w = S_n g$ is a homeomorphism $w: E \to \mathbb{C}$.

It is interesting that in case of fast growth of $M(r,w)$ we find a similar result.

THEOREM 7.2. ($\varepsilon = -1$). Let $w \in \Omega_n'(E)$ and

$$\overline{\lim_{r\to 1-0}} \frac{\log M(r,w)}{\frac{1}{1-r}\log\frac{1}{1-r}} = \infty .$$

Then $w(E)$ contains $\mathbb{C}$ with at most one exceptional value.

It is not known whether exceptional values are actually possible. To prove Theorem 7.2 which holds for the larger class

$$(7.2) \qquad w(z) = \sum_{k=0}^{n} g_k(z) \left(\frac{\bar{z}}{1-z\bar{z}}\right)^k ,$$

where g_k are analytic in E, we make use of a result of Krajkiewicz ([38], Th. 3.2):

THEOREM 3.A. Let f, g be analytic in E. Let f have at most p zeros, and let the functions f, f-g, g have at most m different zeros in E. Let

$$\frac{f(z)}{g(z)} = \sum_{k=-q}^{\infty} a_k z^k , \qquad \mu = \sum_{k=-q}^{p} |a_k| .$$

Then there are constants $0 < \alpha \leq 1 < A$ only depending on m such that for $0 \leq r < R < 1$

$$(7.3) \qquad \log M(r,f) \leq \log M(R,g) + \frac{A}{R-r}[\overset{+}{\log} \frac{\mu}{\alpha^q R^q} + \pi].$$

Proof of Theorem 7.2. Let 0, 1 be exceptional values for the function (7.2) and let

$$g(z,t) = \sum_{k=0}^{n} z^{n-k} g_k(z) \left(\frac{t^2}{1-t^2}\right)^k ,$$

such that $g(z,t) = z^n w(z)$ on $|z| = t$. Each of the functions $g(tz,t)$, $g(tz,t) - (tz)^n$, and $(tz)^n$ has n zeros in E such that the maximal number of their different zeros in E is $m \leq 3n$. For $\mu = \mu(t)$ of Theorem 7.A we find a constant $C(w)$ with

$$\mu(t) \leq C(w)(1-t)^{-n}.$$

(7.3) with $r < t < 1$ gives

$$(7.4) \qquad \log M(r,g(\cdot,t)) \leq \frac{C'(w)}{t-r} \log \frac{1}{1-r} .$$

For z fixed put

$$(7.5) \qquad t_j = 1 - (1-|z|)\left(1 - \frac{j+1}{2n+2}\right), \qquad j = 0, \ldots, n,$$

and solve the equation system

$$g(z,t_j) = \sum_{k=0}^{n} z^{n-k} g_k(z) \left[\frac{t_j^2}{1-t_j^2}\right]^k, \qquad j = 0, \ldots, n,$$

for g_k, $k = 0, \ldots, n$. Then (7.4) gives

$$\log M(r,g_k) \leq \frac{C''(w)}{1-r} \log \frac{1}{1-r} ,$$

and with the common rules for $\overset{+}{\log}$ we obtain

$$\log M(r,w) \leq \sum_{k=0}^{n} \overset{+}{\log} M(r,g_k) + \sum_{k=0}^{n} k \overset{+}{\log} \frac{r}{1-r^2} + \log (n+1)$$

$$\leq \frac{C'''(w)}{1-r} \log \frac{1}{1-r} .$$

The idea of this proof is due to Krajkiewicz, who used it for similar investigations with polyanalytic functions.

7.2. We mention two results on the range of functions in Ω_n, $\varepsilon = 1$.

<u>THEOREM 7.3.</u> ($\varepsilon = 1$). Let G be a simply connected domain in $\mathbb{C}$ with $G \supset \overline{E_{t_n}}$, $w \in \Omega_n'(G)$. Then w has at least one zero in $\overline{E_{t_n}}$.

For the proof we notice that

$$\int_{|z|=t_n} d \arg w(z) \neq 0$$

if $w \neq 0$ for $|z| = t_n$. This follows from the observation that the analytic function with the same boundary values as w vanishes at $z = 0$.

THEOREM 7.4. ($\varepsilon = 1$). Let $w \in \Omega_n(\mathbb{C})$ be bounded. Then there exists a function

$$P(z) = \sum_{k=-n}^{n} a_k z^k,$$

such that $w = S_n P$. Every such function is defined in S^2.

This analogon to Liouville's Theorem follows from an application of Riemann's Theorem on removable singularities and a representation of the functions in $\Omega_n(S^2)$ due to Bauer [4].
For further related results we refer to Chapter 9.

7.3. The possible existence of 'degenerated' mappings is of importance for many investigations. Is it possible that $w \in \Omega_n(G)$ maps G onto an analytic curve? The following result is in [51].

THEOREM 7.5. Let $w \in \Omega_n(G)$ map G onto a differentiable curve. Then there is a $\varphi \in \mathbb{R}$ such that $\tilde{w} = e^{i\varphi} w$ is real valued in G, i.e. the curve is contained in a straight line through the origin.

CHAPTER 8

Uniqueness Theorems

8.1. Since (0.1) is elliptic with analytic coefficients we have the following general uniqueness property: let $w_1, w_2 \in \Omega_n(G)$ and assume $w_1 = w_2$ on an open nonempty subset of G. Then $w_1 \equiv w_2$ in G. Stronger statements are clearly related to the possible location of non-isolated zeros. Here the results are quite similar to those obtained for polyanalytic functions [2]: let $z_o \in G$ be a non-isolated zero of $w \in \Omega_n'(G)$. Then there is an analytic curve C through z_o such that $w = 0$ on $C \cap G$. If w vanishes on those parts of n+1 different circles or straight lines which are passing through G then $w \equiv 0$ in G. Essentially stronger results are not yet available. We consider this a major lack in this theory, and further research should concentrate on this point.
We mention an unusual uniqueness statement using integral means [51].

THEOREM 8.1. Let $w \in \Omega_n'(G)$, $\overline{E_R} \subset G$. Let $\{z_k\}$ be a sequence in E_R converging to $z_o \in E_R$ and assume

$$\int_{|\eta|=R} \frac{\eta^n w(\eta)}{\eta - z_k}\, d\eta = 0, \qquad k \in \mathbb{N}.$$

Then $w = 0$ on $|z| = R$. If $R < t_n$ this implies $w \equiv 0$ in G.

8.2. The results of Chapter 6 enable us to use boundary limits for uniqueness purposes,for instance to obtain an extension of Privaloff's Theorem ([55]). In this context we just mention an extension of an interesting recent result of Dahlberg [18] on harmonic functions.
It is well known that the vanishing of all radial boundary limits of a harmonic function h in a disc is not sufficient to conclude: $h \equiv 0$. For the functions in $\Omega_n(E)$, $\varepsilon = -1$, we have the same phenomenon as seen from the example

$$w_o = S_1\left(\operatorname{Im} \frac{z}{(1-z)^2}\right),$$

which fulfils

$$w_o(re^{i\varphi}) = o((1-r)^{-1}), \qquad \varphi \in \mathbb{R},$$

but $w_o \not\equiv 0$. However, if we pose a uniform restriction on the growth of $M(r,w)$, then such a conclusion is possible.

<u>THEOREM 8.2.</u> ($\varepsilon = -1$). For $w \in \Omega_n(E)$ assume ($r \to 1-0$)

$$\text{i)} \quad w(re^{i\varphi}) = o((1-r)^{-n}), \qquad \varphi \in \mathbb{R},$$

$$\text{ii)} \quad M(r,w) = o((1-r)^{-n-2}).$$

Then $w \equiv 0$.

The case $n = 0$ corresponds to Dahlberg's Theorem.

Proof. Obviously it is enough to consider real valued functions. Let $w = S_n h = \operatorname{Re} S_n g$ where $h = \operatorname{Re} g$, g analytic in E, and let w fulfil i), ii). From i) and Theorem 6.3 we obtain ($r \to 1-0$)

$$h(re^{i\varphi}) = o(1), \qquad \varphi \in \mathbb{R}. \tag{8.1}$$

We have

$$S_n g = \sum_{k=0}^{n} g_k(z) \left[\frac{|z|^2}{1-|z|^2}\right]^{n-k}$$

where

$$g_k(z) = \frac{(2n-k)!}{k!(n-k)!n!} z^{k-n}(z^n g)^{(k)}.$$

Furthermore,

$$\tilde{s}_n(z,t;g) = \sum_{k=0}^{n} g_k(z) \left[\frac{t^2}{1-t^2}\right]^{n-k} \tag{8.2}$$

fulfils

$$\operatorname{Re} \tilde{s}_n(te^{i\varphi},t;g) = w(te^{i\varphi}).$$

We choose t_j, $j = 0, \ldots, n$, as in (7.5) and solve the equation system which arises from (8.2) with $t = t_j$. In particular, for $g_o(z) = \binom{2n}{n} g(z)$ we obtain

$$g_o(z) = \sum_{k=0}^{n} \tilde{s}_n(z,t_k;g) \prod_{\substack{j=0 \\ j \neq k}}^{n} \left[\frac{t_j^2}{1-t_j^2} - \frac{t_k^2}{1-t_k^2}\right]^{-1}.$$

Hence

$$|h(z)| \leq \sum_{k=0}^{n} C_k |\operatorname{Re} \tilde{s}_n(z,t_k;g)| (1-|z|)^n$$

where C_k are certain constants. An application of the maximum principle and ii) gives

$$|\operatorname{Re} \tilde{s}_n(z,t_k;g)| \leq \frac{C(t_k)}{(1-t_k)^{n+2}} \leq \frac{C'(|z|)}{(1-|z|)^{n+2}}$$

with $C'(t) = o(1)$, $t \to 1-0$, and thus

(8.3) $$|h(z)| \leq C''(|z|)(1-|z|)^{-2}, \quad C''(t) = o(1).$$

(8.1) and (8.3) are the assumptions of Dahlberg's Theorem and we conclude $h \equiv 0$ which implies $w \equiv 0$.
It is possible to extend Dahlberg's result also to domains E_R, $R < 1$, instead of E.

CHAPTER 9

Isolated Singularities, Picard's Theorem

9.1. K.W. Bauer and E. Peschl [9] found the representation of single valued solutions of (0.1) in the neighborhood of an isolated singularity.

THEOREM 9.1. Let w be a single valued solution of (0.1) in $U_p(z_o) \subset G_\varepsilon$. Then there exist single valued functions g_1, h_1, analytic in $U_p(z_o)$, and a function

$$P(z) = \sum_{k=-n}^{n} a_k z^k ,$$

such that $w = S_n(g+\bar{h})$ in $U_p(z_o)$, where

$$g(z) = g_1(z) + P(z) \log (z-z_o),$$

$$h(z) = h_1(z) + \overline{P\left(\frac{-\varepsilon}{\bar{z}}\right)} \log (z-z_o).$$

For $w \in \Omega_n'(U_p(z_o))$ we have $w = S_n g$ for a single valued analytic function g in $U_p(z_o)$.

Other results in [9] are dealing with real valued solutions and their singularities and with solutions which have only singularities of logarithmic type. In extension of these results, solutions with multiplicative behaviour in $U_p(z_o)$ have been studied by L. Reich [48].

The growth of solutions near isolated singularities has no specific properties beside those which are already valid for the more general equations studied by Bers [10]: removable singularities and behaviour near pole-type singularities.
In [55] we established an extension of the Big Picard Theorem [1] for $w \in \Omega_n''(U_p(z_o))$ with an essential singularity at z_o. In the sequel we

---

[1] Chernoff [17] proves the Small Picard Theorem for entire functions representable through certain Bergman operators, which is also applicable to (0.1).

shall show that better results can be obtained using the theory of multianalytic functions. In view of the invariance of (0.1) and all other relevant entities w.r.t. the corresponding geometry we restrict ourselves to $z_o = 0$.

Following Krajkiewicz [37], the point $z = 0$ is an essential singularity of a multianalytic function

$$w = \sum_{k=-m}^{\infty} g_k(z)\bar{z}^k ,$$

if for the orders d_k of the singularities of $z^k g_k(z)$ at $z = 0$ we have

$$\sup_k d_k = \infty .$$

An 'exceptional value' of a singular multianalytic function is defined to be a multianalytic function $\tilde{w}$ without essential singularity at $z = 0$ such that $w - \tilde{w} \neq 0$ in a punctured neighborhood of $z = 0$. Two exceptional values are different if they are different in a punctured neighborhood of $z = 0$. Note that constants are candidates for exceptional values.

THEOREM 9.A. [37] A multivalent function which has an essential singularity at $z = 0$ has at most one exceptional value.

From Theorem 1.5 and the representation (1.30) we conclude: $w = S_n g$, g analytic in $U_p(0)$, is multianalytic in $U_p(0)$ with an essential singularity at $z = 0$ if and only if g has an essential singularity at $z = 0$.

THEOREM 9.2. Let g be analytic in $U_p(0)$ with an essential singularity at $z = 0$. Then there is at most one harmonic function $f = f_1 + \overline{f_2}$ in $U_p(0)$, f_j analytic and not essentially singular at $z = 0$, such that $w = S_n(g+f)$ omits a complex number as value in $U_p(0)$.

Exceptional values actually exist as shown by the function

$$S_1(z^{-1}e^{1/z})$$

which is different from zero in a small neighborhood of the origin.

A consequence of Picard's Theorem is the following extension of

Schottky's Theorem, see [55].

THEOREM 9.3. Let $a,p,q \in \mathbb{C}$, $p \neq q$. Let $0 < R' < R$. Then there exists $M = M(a,p,q,R',R)$ such that for all $w \in \Omega_n'(E_R)$ with $w(0) = a$ and $w \neq p,q$ in E_R we have: $|w(z)| < M$, $z \in E_R$.

9.2. The existence of two linear independent solutions in Ω_n which depend only on $|z|$, namely $S_n(1)$ and $S_n(\log |z|)$ combined with the maximum principle makes it easy to obtain estimates for solutions defined in an annulus $K = \{ z | 0 < R_1 < |z| < R_2 \}$, $R_2 < t_n'$: a generalization of Hadamard's Three Circle Theorem.

THEOREM 9.4. Let $w \in \Omega_n(K)$ and

$$\overline{\lim_{r \to R_j}} M(r,w) = M_j, \qquad j = 1,2.$$

Then, for $R_1 < |z| < R_2$

$$|w(z)| \leq f(|z|) = \alpha S_n(|z|^0) + \beta S_n(\log |z|),$$

where the constants α, β are determined by the relation $f(R_j) = M_j$, $j = 1,2$.

9.3. For solutions of (0.1), $\varepsilon = -1$, in E with finitely many logarithmic singularities and a certain behaviour on the boundary, sharp estimates have been obtained by K.W. Bauer. These strong and difficult results are too complicated to be reproduced here. We refer to the original paper [3].

CHAPTER 10

Analytic continuation

10.1. In this section we study solutions of (0.1), $\varepsilon = -1$, which can be continued across the singular unit circle. The details are found in [52].

Let B_n be class of solutions of (0.1), $\varepsilon = -1$, such that

i) they are defined in E upto isolated singularities,

ii) they remain uniformly bounded if approaching ∂E.

An important member of B_n is $S_n(\log |z|)$ and from Theorem 9.4 we derive for $w \in B_n$:

1) $M(r,w)/S_n(\log 1/r)$ is monotonic decreasing for $R_o < r < 1$ if w has no singularities in $R_o < |z| < 1$.

2) $M(r,w) = O((1-r)^{n+1})$, $r \to 1-0$.

The following representation is obtained by subtraction of suitable singularity functions and an application of Riemann's Theorem on removable singularities.

THEOREM 10.1. ($\varepsilon = -1$). $w \in B_n$ if and only if

$$w = S_n(f(z) - f(1/\bar{z})),$$

where

$$f(z) = g(z) + \sum_{k=1}^{m} P_k(z) \log \frac{z-z_k}{1-\bar{z}_k z}$$

with the following restrictions:

i) g(z) is a single valued analytic function in S^2 without singularities on ∂E.

ii) $z_k \in E$, $k = 1, \ldots, m$,

iii) $P_k(z) = \sum_{j=-n}^{n} a_{jk} z^k$, $k = 1, \ldots, m$.

This result contains the following further properties of B_n:

3) $w \in B_n$ has a real analytic continuation across ∂E by means of the formula

$$w(z) = (-1)^{n+1} w(1/\bar{z}).$$

4) If $w = S_n g \in B_n$ where g is single valued and analytic upto isolated singularities in E, then $w \equiv 0$.

Similar results as those of Theorem 10.1 have been applied by K.W. Bauer [6] and G. Jank [33] to the construction of global fundamental solutions of related differential equations.

10.2. A general reflection principle, applicable also to (0.1) is due to H. Lewy [39]. However, differentiability on the line of reflection is required in his results. As in the case of analytic functions this is not always necessary for (0.1), as shown by T. Rüdiger [50]. His proof makes use of an extension of the Poisson formula for semi circles. For a similar approach in the analytic case see Dinghas [19].

THEOREM 10.2. Let $w \in \Omega_n'(E_R \cap \{z \mid \mathrm{Re}\, z > 0\})$, $R < t_n'$, and assume $\mathrm{Im}\, w \to 0$ whenever z approaches the real axis. Then w can be continued into a function in $\Omega_n(E_R)$ and this continuation fulfils

$$w(z) = \overline{w(\bar{z})}, \quad z \in E_R \cap \{z \mid \mathrm{Re}\, z < 0\} .$$

It is not unlikely that similar results hold for other geometrical configurations also.

CHAPTER 11

Automorphic Solutions

11.1. Equations (0.1) and (0.2) are closely related to the theory of automorphic functions and automorphic forms, see the introduction. Bauer's representation for the solutions of (0.1) contributs a new aspect to this theory. We shall characterize the generating functions of automorphic solutions in Bauer's representation. Again, the parameters $n \in \mathbb{N}$ in (0.1) appear to be distinguished.

Let γ be a subgroup of the conformal automorphism of E. $A_n(\gamma)$ is the set of solutions of (0.1), $\varepsilon = -1$, such that

i) they are defined in E upto isolated singularities which have no logarithmic part,

ii) $w(z) = w(a(z))$, $z \in E$, for every $a \in \gamma$.

A singular Eichler integral of order n is a function g defined in S^2 and analytic in both, E and $S^2 \setminus \bar{E}$, upto isolated singularities, such that for every $a \in \gamma$

$$g(z) - a'(z)^{-n} g(a(z))$$

is the restriction of a polynomial of degree $\leq 2n$ to $S^2 \setminus \partial E$.

THEOREM 11.1. ($\varepsilon = -1$). $w \in A_n(\gamma)$ if and only if there is a singular Eichler integral g of order n w.r.t. γ such that for $f(z) = z^{-n} g(z)$ we have

$$w = S_n(g(z) - g(1/\bar{z})), \quad z \in E.$$

This result, proved for (0.2), is in [25], and in the case of finitely generated horocyclic groups all members of $A_n(\gamma)$ have been constructed by the means of certain Eisenstein series and Theorem 11.1. The interaction of the vast theory of Eichler integrals (see Bers [11]) and the function theory of the solutions of (0.1) proved to be fruitful: beside the rather complete results on $A_n(\gamma)$ some minor problems on Eichler integrals have been solved in this context. For details see [25], [54], [32], [42].

11.2. A different construction principle for members of $A_n(\gamma)$ has been given by E. Peschl (see Bauer [4]). It makes use of the absolute invariants

$$\alpha_2 = \frac{1}{2}\frac{f''}{f'^2} - \frac{1}{f'}\frac{\bar{z}}{1-z\bar{z}} ,$$

$$\beta_3 = \frac{1}{6f'^2}\left[\left(\frac{f''}{f'}\right)' - \frac{1}{2}\left(\frac{f''}{f'}\right)^2\right] ,$$

$$\beta_n = \frac{1}{nf'}\frac{\partial}{\partial z}\beta_{n-1} , \quad n \geq 4,$$

and represents members of $A_n(\gamma)$ as polynomials in α_2 with certain coefficients which are analytic in β_k , where f is an arbitrary meromorphic automorphic function in E. It is not possible, however, to represent every function in $A_n(\gamma)$ by this method.

11.3. For certain continuous subgroups γ automorphic solutions of (0.1), $\varepsilon = -1$, have been determined by K.W. Bauer [4]. For instance, all solutions of (0.1), $\varepsilon = -1$, which are constant on all circles in E, touching E in $z = 1$, are given by

$$w = C_1\left[\frac{|z-1|^2}{1-|z|^2}\right]^n + C_2\left[\frac{|z-1|^2}{1-|z|^2}\right]^{-n-1}, \quad C_1, C_2 \in \mathbb{C}.$$

And all solutions constant on the circles through $z = \pm 1$ can be represented by

$$w = C_1 P_n\left(\frac{\bar{z}-z}{1-z\bar{z}}\right) + C_2 Q_n\left(\frac{\bar{z}-z}{1-z\bar{z}}\right) , \quad C_1, C_2 \in \mathbb{C}.$$

For further details see [4].

REFERENCES

[1] M. Abramowitz, I.A. Stegun, Handbook of Mathematical Functions. Dover Publications, New York 1965.

[2] M.B. Balk, M.F. Zuev, On Polyanalytic Functions, Uspeki Mat. Nauk., 25, 203-226 (1970). (Engl. Translation: Soviet Math. Dokl. 12, 61-65 (1971).)

[3] K.W. Bauer, Über die Lösungen der elliptischen Differentialgleichung $(1\pm z\bar{z})^2 w_{z\bar{z}} + \lambda w = 0$, J. reine angew. Math. 221, Teil I: 48-84, Teil II: 176-196 (1966).

[4] ----, Über eine der Differentialgleichung $(1\pm z\bar{z})^2 w_{z\bar{z}} \pm n(n+1)w=0$ zugeordnete Funktionentheorie, Bonner Math. Schr., 23 (1965).

[5] ----, Zur Verallgemeinerung des Poissonschen Satzes, Ann. Akad. Sci. Fenn., Ser. A., 437, 1-28 (1968).

[6] ----, Differentialoperatoren bei einer Klasse verallgemeinerter Tricomi-Gleichungen, ZAMM, 54, 715-721 (1974).

[7] ----, Eine Darstellung der allgemeinen Kugelfunktionen, Ber. d. Ges. f. Math. u. Datenv. Bonn, 57, 5-11 (1972).

[8] ----, Polynomoperatoren bei Differentialgleichungen der Form $w_{z\bar{z}} + Aw_{\bar{z}} + Bw = 0$, J. reine angew. Math., 283/284, 364-369 (1976).

[9] ----, E. Peschl, Ein allgemeiner Entwicklungssatz für die Lösungen der Differentialgleichung $(1+\varepsilon z\bar{z})^2 w_{z\bar{z}} + \varepsilon n(n+1)w = 0$ in der Nähe isolierter Singularitäten, S.-ber. der Bayer. Akademie d. Wiss., math.-nat. Klasse, 113-146 (1965).

[10] L. Bers, Local Behaviour of Solutions of General Linear Elliptic Equations, Comm. Pure Appl. Math., 8, 473-496 (1955).

[11] ----, Eichler Integrals with Singularities, Acta Math., 127, 11-22 (1971).

[12] R.P. Boas, Asymptotic Relations for Derivatives, Duke Math. Journ., 3, 637-646 (1937).

[13] D. Borwein, On a Scale of Abel-Type Summability Methods, Proc. Cambridge Phil. Soc., 53, 318-322 (1957).

[14] ----, J.H. Rizvi, On Abel-Type Methods of Summability, J. reine angew. Math., 247, 139-145 (1971).

[15] L. Brickman, Subordinate Families of Analytic Functions, Illinois J. Math., 15, 241-248 (1971).

[16] P.L. Butzer, R.J. Nessel, Fourier-Analysis and Approximation, Birkhäuser Verlag 1971.

[17] H. Chernoff, Complex Solutions of Partial Differential Equations,

Amer. J. Math., 68, 455-478 (1946).

[18] B. Dahlberg, Representation of Harmonic Functions in the Unit Disc, Preprint Univ. Goeteborg.

[19] A. Dinghas, Vorlesungen über Funktionentheorie, Springer 1961.

[20] P.J. Eenigenburg, F.R. Keogh, On the Hardy Class of some Univalent Functions and Their Derivatives, Mich. Math. J., 17, 335-346 (1970).

[21] J. Elstrodt, W. Roelcke, Über das wesentliche Spektrum zum Eigenwertproblem der automorphen Formen, Manuscripta Math., 11, 391-406 (1974).

[22] G. Frank, Eine Vermutung von Hayman über Nullstellen meromorpher Funktionen, Math. Z., 149, 29-36 (1976).

[23] M.P. Ganin, Das Dirichlet Problem für die Gleichung $\Delta u + [4n(n+1)/(1+x^2+y^2)^2]u = 0$. Uspeki Mat. Nauk., 12-5 (77), 205-209 (1957).

[24] R.P. Gilbert, Function Theoretic Methods in Partial Differential Equations. Academic Press 1969.

[25] I. Haeseler, St. Ruscheweyh, Singuläre Eichlerintegrale und verallgemeinerte Eisensteinreihen, Math. Ann., 203, 251-259 (1973).

[26] G.H. Hardy, Generalizations of a limit Theorem of Mr. Mercer, Quart. J. pure appl. Math., 43, 143-150 (1912).

[27] R. Heersink, Characterization of Certain Differential Operators in the Solution of Linear Partial Differential Equations, Glasgow Math. Journal, 17,2, 83-88 (1976).

[28] ----, Lösungsdarstellungen mittels Differentialoperatoren für das Dirichlet Problem der Gleichung $\Delta u+c(x,y)u = 0$, Lect. Notes in Math., 561, 227-238 (1976).

[29] ----, Über Lösungsdarstellungen und funktionentheoretische Methoden bei elliptischen Differentialgleichungen, Ber. d. Forschungszentrum Graz, 67 (1976).

[30] A. Huber, On uniqueness of generalized axially symmetric potentials, Ann. of Math., 60, 351-385 (1954).

[31] ----, Some Results on Generalized Axially Symmetric Potential, Proc. Conf. Part. Diff. Equ., Maryland, 147-155 (1955).

[32] G. Jank, Automorphe Lösungen der Euler-Darboux Gleichung, Lect. Notes in Math., 561, 277-282 (1976).

[33] ----, Funktionentheoretische Untersuchungen von Lösungen gewisser elliptischer Differentialgleichungen, Preprint.

[34] ----, St. Ruscheweyh, Funktionenfamilien mit einem Maximumprinzip und elliptische Differentialgleichungen II, Monatsh.

Math., 79, 103-113, (1975).

[35] K. Knopp, Theorie und Anwendung der unendlichen Reihen, Springer 1965.

[36] M. Kracht, E. Kreyszig, Bergman-Operatoren mit Polynomen als Erzeugenden, Manus. Math., 1, 369-376 (1969).

[37] P. Krajkiewicz, The Picard Theorem for Multianalytic Functions, Pacific J. Math., 48, 423-439 (1973).

[38] ----, A new proof of a theorem of Saxer, Ann. Pol. Math., 33, 293-309 (1977).

[39] H. Lewy, On the reflection laws of second order differential equations in two independent variables, Bull. Amer. Math. Soc., 65, 37-58 (1959).

[40] H. Maaβ, Über eine neue Art von nichtanalytischen automorphen Funktionen und die Bestimmung Dirichletscher Reihen durch Funktionalgleichungen, Math. Ann., 121, 141-183 (1949).

[41] T.H. MacGregor, Certain Integrals of univalent and convex functions, Math. Z., 103, 48-54 (1968).

[42] E. Meister, Dipl. Arbeit Univ. Heidelberg 1956.

[43] B.P. Mishra, Strong summability of infinite series on a scale of Abel-type summability methods, Proc. Cambr. Phil. Soc., 63, 119-127 (1967).

[44] G. Pólya, I.J. Schoenberg, Remarks on de la Vallée Poussin means and convex conformal maps of the circle, Pac. J. Math., 8, 295-333 (1958).

[45] Ch. Pommerenke, Über die Subordination analytischer Funktionen, J. reine angew. Math., 218, 159-173 (1965).

[46] H. Prawitz, Über Mittelwerte analytischer Funktionen, Arkiv Mat. Astr. Fys., 20, 1-12 (1927).

[47] M.H. Protter, H.F. Weinberger, Maximum principles in Differential Equations, Prentice Hall, Englewood Cliffs, 1967.

[48] L. Reich, Über multiplikative und algebraisch verzweigte Lösungen der Differentialgleichung $(1+\varepsilon z\bar{z})^2 w_{z\bar{z}} + \varepsilon n(n+1)w = 0$, Ber. d. Ges. f. Math. u. Datenv., Bonn, 57, 13-28 (1972).

[49] W. Roelcke, Über die Wellengleichung bei Grenzkreisgruppen 1. Art, S.-ber. Heidelberger Akad. Wiss., math.-naturw. Kl. 1953-55, 159-267 (1956).

[50] T. Rüdiger, Spiegelungsprinzipien bei partiellen Differentialgleichungen, Diplomarbeit Bonn 1973.

[51] St. Ruscheweyh, Gewisse Klassen verallgemeinerter analytischer Funktionen, Bonner math. Schr. 39 (1970).

[52] ----, Über den Rand des Einheitskreises hinaus fortsetzbare Lö-

sungen der Differentialgleichung von Peschl und Bauer, Ber. d. Ges. f. Math. u. Datenv., Bonn, 57, 29-36 (1972).

[53] ----, Eine Verallgemeinerung der Leibnizschen Produktregel, Math. Nachr., 58, 241-245 (1973).

[54] ----, Beschränkte Eichlerintegrale, Ber. d. Ges. f. Math. u. Datenv., Bonn, 77, 143-145 (1973).

[55] ----, Geometrische Eigenschaften der Lösungen der Differentialgleichung $(1-z\bar{z})^2 w_{z\bar{z}} - n(n+1)w = 0$, J. reine angew. Math., 270, 143-157 (1974).

[56] ----, Hardy spaces of λ-harmonic functions, Research Notes in Math., Pitmans Publ., 8, 68-84 (1976).

[57] ----, On the mapping problem for certain second order elliptic equations, Lect. Notes in Math., 561, 421-429 (1976).

[58] ----, Funktionentheoretische Methoden bei partiellen Differentialgleichungen, Ber. Forschungszentrum Graz, 60 (1976).

[59] ----, K. J. Wirths, Mittelwerteigenschaften bei selbstadjungierten elliptischen Differentialgleichungen, Ber. Ges. Math. Datenv., Bonn, 75, 21-24 (1973).

[60] ----, K.J. Wirths, Riemann's mapping theorem for n-analytic functions, Math. Z., 149, 287-297 (1976).

[61] ----, T. Sheil-Small, Hadamard Products of schlicht functions and the Pólya-Schoenberg conjecture, Comm. Math. Helv., 48, 119-135 (1973).

[62] ----, Perfect summability methods generated by partial differential equations, Preprint.

[63] T. Sheil-Small, On the convolution of analytic functions, J. reine angew. Math., 258, 137-152 (1973).

[64] I.N. Vekua, New methods for solving elliptic equations, North Holland Publ. 1968.

[65] J.J. Walker, Theorems in the calculus of operations, Proc. London Math. Soc. 11, 108-113 (1880).

[66] A. Weinstein, Generalized axially symmetric potential theory, Bull. Amer. Math. Soc., 59, 20-83 (1953).

[67] E.T. Whittaker, G.M. Watson, Modern analysis, 4. ed., Cambr. Univ. Press 1927.

[68] L. Włodarski, Sur les methodes continues limitation (I), Studia Math., 14, 161-187 (1954).

SUBJECT INDEX

GLOSSARY

$\bar{A}$	closure of A
$\dot{A}$	interior of A
∂A	boundary of A
clco	closed convex hull of A
E	unit disc
E_R	$\{z \mid \lvert z\rvert < R\}$
G_ε	195
S^2	Riemannian sphere
$U_p(0)$	203
$\mathbb{N}$	natural numbers
$\mathbb{N}_0$	$\mathbb{N} \cup \{0\}$
$\mathbb{R}$	real numbers
$\mathbb{C}$	complex numbers
Ω_n	195
Ω_n'	197
Ω_n''	197
Λ_n	197
Λ_n'	197
B_n	246
$A_n(\gamma)$	248
h_n^p	227
H_n^p	228
E_n	195
S_n	196
F_n	198
Δ^n	196
Re	real part
Im	imaginary part
$\dot{f}$	220
$\bar{z}$	complex conjugate
$\prec$	219
$M(r,w)$	213

Vol. 640: J. L. Dupont, Curvature and Characteristic Classes. X, 175 pages. 1978.

Vol. 641: Séminaire d'Algèbre Paul Dubreil, Proceedings Paris 1976–1977. Edité par M. P. Malliavin. IV, 367 pages. 1978.

Vol. 642: Theory and Applications of Graphs, Proceedings, Michigan 1976. Edited by Y. Alavi and D. R. Lick. XIV, 635 pages. 1978.

Vol. 643: M. Davis, Multiaxial Actions on Manifolds. VI, 141 pages. 1978.

Vol. 644: Vector Space Measures and Applications I, Proceedings 1977. Edited by R. M. Aron and S. Dineen. VIII, 451 pages. 1978.

Vol. 645: Vector Space Measures and Applications II, Proceedings 1977. Edited by R. M. Aron and S. Dineen. VIII, 218 pages. 1978.

Vol. 646: O. Tammi, Extremum Problems for Bounded Univalent Functions. VIII, 313 pages. 1978.

Vol. 647: L. J. Ratliff, Jr., Chain Conjectures in Ring Theory. VIII, 133 pages. 1978.

Vol. 648: Nonlinear Partial Differential Equations and Applications, Proceedings, Indiana 1976–1977. Edited by J. M. Chadam. VI, 206 pages. 1978.

Vol. 649: Séminaire de Probabilités XII, Proceedings, Strasbourg, 1976–1977. Edité par C. Dellacherie, P. A. Meyer et M. Weil. VIII, 805 pages. 1978.

Vol. 650: C*-Algebras and Applications to Physics. Proceedings 1977. Edited by H. Araki and R. V. Kadison. V, 192 pages. 1978.

Vol. 651: P. W. Michor, Functors and Categories of Banach Spaces. VI, 99 pages. 1978.

Vol. 652: Differential Topology, Foliations and Gelfand-Fuks-Cohomology, Proceedings 1976. Edited by P. A. Schweitzer. XIV, 252 pages. 1978.

Vol. 653: Locally Interacting Systems and Their Application in Biology. Proceedings, 1976. Edited by R. L. Dobrushin, V. I. Kryukov and A. L. Toom. XI, 202 pages. 1978.

Vol. 654: J. P. Buhler, Icosahedral Golois Representations. III, 143 pages. 1978.

Vol. 655: R. Baeza, Quadratic Forms Over Semilocal Rings. VI, 199 pages. 1978.

Vol. 656: Probability Theory on Vector Spaces. Proceedings, 1977. Edited by A. Weron. VIII, 274 pages. 1978.

Vol. 657: Geometric Applications of Homotopy Theory I, Proceedings 1977. Edited by M. G. Barratt and M. E. Mahowald. VIII, 459 pages. 1978.

Vol. 658: Geometric Applications of Homotopy Theory II, Proceedings 1977. Edited by M. G. Barratt and M. E. Mahowald. VIII, 487 pages. 1978.

Vol. 659: Bruckner, Differentiation of Real Functions. X, 247 pages. 1978.

Vol. 660: Equations aux Dérivée Partielles. Proceedings, 1977. Edité par Pham The Lai. VI, 216 pages. 1978.

Vol. 661: P. T. Johnstone, R. Paré, R. D. Rosebrugh, D. Schumacher, R. J. Wood, and G. C. Wraith, Indexed Categories and Their Applications. VII, 260 pages. 1978.

Vol. 662: Akin, The Metric Theory of Banach Manifolds. XIX, 306 pages. 1978.

Vol. 663: J. F. Berglund, H. D. Junghenn, P. Milnes, Compact Right Topological Semigroups and Generalizations of Almost Periodicity. X, 243 pages. 1978.

Vol. 664: Algebraic and Geometric Topology, Proceedings, 1977. Edited by K. C. Millett. XI, 240 pages. 1978.

Vol. 665: Journées d'Analyse Non Linéaire. Proceedings, 1977. Edité par P. Bénilan et J. Robert. VIII, 256 pages. 1978.

Vol. 666: B. Beauzamy, Espaces d'Interpolation Réels: Topologie et Géometrie. X, 104 pages. 1978.

Vol. 667: J. Gilewicz, Approximants de Padé. XIV, 511 pages. 1978.

Vol. 668: The Structure of Attractors in Dynamical Systems. Proceedings, 1977. Edited by J. C. Martin, N. G. Markley and W. Perrizo. VI, 264 pages. 1978.

Vol. 669: Higher Set Theory. Proceedings, 1977. Edited by G. H. Müller and D. S. Scott. XII, 476 pages. 1978.

Vol. 670: Fonctions de Plusieurs Variables Complexes III, Proceedings, 1977. Edité par F. Norguet. XII, 394 pages. 1978.

Vol. 671: R. T. Smythe and J. C. Wierman, First-Passage Perculation on the Square Lattice. VIII, 196 pages. 1978.

Vol. 672: R. L. Taylor, Stochastic Convergence of Weighted Sums of Random Elements in Linear Spaces. VII, 216 pages. 1978.

Vol. 673: Algebraic Topology, Proceedings 1977. Edited by P. Hoffman, R. Piccinini and D. Sjerve. VI, 278 pages. 1978.

Vol. 674: Z. Fiedorowicz and S. Priddy, Homology of Classical Groups Over Finite Fields and Their Associated Infinite Loop Spaces. VI, 434 pages. 1978.

Vol. 675: J. Galambos and S. Kotz, Characterizations of Probability Distributions. VIII, 169 pages. 1978.

Vol. 676: Differential Geometrical Methods in Mathematical Physics II, Proceedings, 1977. Edited by K. Bleuler, H. R. Petry and A. Reetz. VI, 626 pages. 1978.

Vol. 677: Séminaire Bourbaki, vol. 1976/77, Exposés 489–506. IV, 264 pages. 1978.

Vol. 678: D. Dacunha-Castelle, H. Heyer et B. Roynette. Ecole d'Eté de Probabilités de Saint-Flour. VII-1977. Edité par P. L. Hennequin. IX, 379 pages. 1978.

Vol. 679: Numerical Treatment of Differential Equations in Applications, Proceedings, 1977. Edited by R. Ansorge and W. Törnig. IX, 163 pages. 1978.

Vol. 680: Mathematical Control Theory, Proceedings, 1977. Edited by W. A. Coppel. IX, 257 pages. 1978.

Vol. 681: Séminaire de Théorie du Potentiel Paris, No. 3, Directeurs: M. Brelot, G. Choquet et J. Deny. Rédacteurs: F. Hirsch et G. Mokobodzki. VII, 294 pages. 1978.

Vol. 682: G. D. James, The Representation Theory of the Symmetric Groups. V, 156 pages. 1978.

Vol. 683: Variétés Analytiques Compactes, Proceedings, 1977. Edité par Y. Hervier et A. Hirschowitz. V, 248 pages. 1978.

Vol. 684: E. E. Rosinger, Distributions and Nonlinear Partial Differential Equations. XI, 146 pages. 1978.

Vol. 685: Knot Theory, Proceedings, 1977. Edited by J. C. Hausmann. VII, 311 pages. 1978.

Vol. 686: Combinatorial Mathematics, Proceedings, 1977. Edited by D. A. Holton and J. Seberry. IX, 353 pages. 1978.

Vol. 687: Algebraic Geometry, Proceedings, 1977. Edited by L. D. Olson. V, 244 pages. 1978.

Vol. 688: J. Dydak and J. Segal, Shape Theory. VI, 150 pages. 1978.

Vol. 689: Cabal Seminar 76–77, Proceedings, 1976–77. Edited by A.S. Kechris and Y. N. Moschovakis. V, 282 pages. 1978.

Vol. 690: W. J. J. Rey, Robust Statistical Methods. VI, 128 pages. 1978.

Vol. 691: G. Viennot, Algèbres de Lie Libres et Monoïdes Libres. III, 124 pages. 1978.

Vol. 692: T. Husain and S. M. Khaleelulla, Barrelledness in Topological and Ordered Vector Spaces. IX, 258 pages. 1978.

Vol. 693: Hilbert Space Operators, Proceedings, 1977. Edited by J. M. Bachar Jr. and D. W. Hadwin. VIII, 184 pages. 1978.

Vol. 694: Séminaire Pierre Lelong – Henri Skoda (Analyse) Année 1976/77. VII, 334 pages. 1978.

Vol. 695: Measure Theory Applications to Stochastic Analysis, Proceedings, 1977. Edited by G. Kallianpur and D. Kölzow. XII, 261 pages. 1978.

Vol. 696: P. J. Feinsilver, Special Functions, Probability Semigroups, and Hamiltonian Flows. VI, 112 pages. 1978.

Vol. 697: Topics in Algebra, Proceedings, 1978. Edited by M. F. Newman. XI, 229 pages. 1978.

Vol. 698: E. Grosswald, Bessel Polynomials. XIV, 182 pages. 1978.

Vol. 699: R. E. Greene and H.-H. Wu, Function Theory on Manifolds Which Possess a Pole. III, 215 pages. 1979.

Vol. 700: Module Theory, Proceedings, 1977. Edited by C. Faith and S. Wiegand. X, 239 pages. 1979.

Vol. 701: Functional Analysis Methods in Numerical Analysis, Proceedings, 1977. Edited by M. Zuhair Nashed. VII, 333 pages. 1979.

Vol. 702: Yuri N. Bibikov, Local Theory of Nonlinear Analytic Ordinary Differential Equations. IX, 147 pages. 1979.

Vol. 703: Equadiff IV, Proceedings, 1977. Edited by J. Fábera. XIX, 441 pages. 1979.

Vol. 704: Computing Methods in Applied Sciences and Engineering, 1977, I. Proceedings, 1977. Edited by R. Glowinski and J. L. Lions. VI, 391 pages. 1979.

Vol. 705: O. Forster und K. Knorr, Konstruktion verseller Familien kompakter komplexer Räume. VII, 141 Seiten. 1979.

Vol. 706: Probability Measures on Groups, Proceedings, 1978. Edited by H. Heyer. XIII, 348 pages. 1979.

Vol. 707: R. Zielke, Discontinuous Čebyšev Systems. VI, 111 pages. 1979.

Vol. 708: J. P. Jouanolou, Equations de Pfaff algébriques. V, 255 pages. 1979.

Vol. 709: Probability in Banach Spaces II. Proceedings, 1978. Edited by A. Beck. V, 205 pages. 1979.

Vol. 710: Séminaire Bourbaki vol. 1977/78, Exposés 507-524. IV, 328 pages. 1979.

Vol. 711: Asymptotic Analysis. Edited by F. Verhulst. V, 240 pages. 1979.

Vol. 712: Equations Différentielles et Systèmes de Pfaff dans le Champ Complexe. Edité par R. Gérard et J.-P. Ramis. V, 364 pages. 1979.

Vol. 713: Séminaire de Théorie du Potentiel, Paris No. 4. Edité par F. Hirsch et G. Mokobodzki. VII, 281 pages. 1979.

Vol. 714: J. Jacod, Calcul Stochastique et Problèmes de Martingales. X, 539 pages. 1979.

Vol. 715: Inder Bir S. Passi, Group Rings and Their Augmentation Ideals. VI, 137 pages. 1979.

Vol. 716: M. A. Scheunert, The Theory of Lie Superalgebras. X, 271 pages. 1979.

Vol. 717: Grosser, Bidualräume und Vervollständigungen von Banachmoduln. III, 209 pages. 1979.

Vol. 718: J. Ferrante and C. W. Rackoff, The Computational Complexity of Logical Theories. X, 243 pages. 1979.

Vol. 719: Categorial Topology, Proceedings, 1978. Edited by H. Herrlich and G. Preuß. XII, 420 pages. 1979.

Vol. 720: E. Dubinsky, The Structure of Nuclear Fréchet Spaces. V, 187 pages. 1979.

Vol. 721: Séminaire de Probabilités XIII. Proceedings, Strasbourg, 1977/78. Edité par C. Dellacherie, P. A. Meyer et M. Weil. VII, 647 pages. 1979.

Vol. 722: Topology of Low-Dimensional Manifolds. Proceedings, 1977. Edited by R. Fenn. VI, 154 pages. 1979.

Vol. 723: W. Brandal, Commutative Rings whose Finitely Generated Modules Decompose. II, 116 pages. 1979.

Vol. 724: D. Griffeath, Additive and Cancellative Interacting Particle Systems. V, 108 pages. 1979.

Vol. 725: Algèbres d'Opérateurs. Proceedings, 1978. Edité par P. de la Harpe. VII, 309 pages. 1979.

Vol. 726: Y.-C. Wong, Schwartz Spaces, Nuclear Spaces and Tensor Products. VI, 418 pages. 1979.

Vol. 727: Y. Saito, Spectral Representations for Schrödinger Operators With Long-Range Potentials. V, 149 pages. 1979.

Vol. 728: Non-Commutative Harmonic Analysis. Proceedings, 1978. Edited by J. Carmona and M. Vergne. V, 244 pages. 1979.

Vol. 729: Ergodic Theory. Proceedings, 1978. Edited by M. Denker and K. Jacobs. XII, 209 pages. 1979.

Vol. 730: Functional Differential Equations and Approximation of Fixed Points. Proceedings, 1978. Edited by H.-O. Peitgen and H.-O. Walther. XV, 503 pages. 1979.

Vol. 731: Y. Nakagami and M. Takesaki, Duality for Crossed Products of von Neumann Algebras. IX, 139 pages. 1979.

Vol. 732: Algebraic Geometry. Proceedings, 1978. Edited by K. Lønsted. IV, 658 pages. 1979.

Vol. 733: F. Bloom, Modern Differential Geometric Techniques in the Theory of Continuous Distributions of Dislocations. XII, 206 pages. 1979.

Vol. 734: Ring Theory, Waterloo, 1978. Proceedings, 1978. Edited by D. Handelman and J. Lawrence. XI, 352 pages. 1979.

Vol. 735: B. Aupetit, Propriétés Spectrales des Algèbres de Banach. XII, 192 pages. 1979.

Vol. 736: E. Behrends, M-Structure and the Banach-Stone Theorem. X, 217 pages. 1979.

Vol. 737: Volterra Equations. Proceedings 1978. Edited by S.-O. Londen and O. J. Staffans. VIII, 314 pages. 1979.

Vol. 738: P. E. Conner, Differentiable Periodic Maps. 2nd edition, IV, 181 pages. 1979.

Vol. 739: Analyse Harmonique sur les Groupes de Lie II. Proceedings, 1976-78. Edited by P. Eymard et al. VI, 646 pages. 1979.

Vol. 740: Séminaire d'Algèbre Paul Dubreil. Proceedings, 1977-78. Edited by M.-P. Malliavin. V, 456 pages. 1979.

Vol. 741: Algebraic Topology, Waterloo 1978. Proceedings. Edited by P. Hoffman and V. Snaith. XI, 655 pages. 1979.

Vol. 742: K. Clancey, Seminormal Operators. VII, 125 pages. 1979.

Vol. 743: Romanian-Finnish Seminar on Complex Analysis. Proceedings, 1976. Edited by C. Andreian Cazacu et al. XVI, 713 pages. 1979.

Vol. 744: I. Reiner and K. W. Roggenkamp, Integral Representations. VIII, 275 pages. 1979.

Vol. 745: D. K. Haley, Equational Compactness in Rings. III, 167 pages. 1979.

Vol. 746: P. Hoffman, τ-Rings and Wreath Product Representations. V, 148 pages. 1979.

Vol. 747: Complex Analysis, Joensuu 1978. Proceedings, 1978. Edited by I. Laine, O. Lehto and T. Sorvali. XV, 450 pages. 1979.

Vol. 748: Combinatorial Mathematics VI. Proceedings, 1978. Edited by A. F. Horadam and W. D. Wallis. IX, 206 pages. 1979.

Vol. 749: V. Girault and P.-A. Raviart, Finite Element Approximation of the Navier-Stokes Equations. VII, 200 pages. 1979.

Vol. 750: J. C. Jantzen, Moduln mit einem höchsten Gewicht. III, 195 Seiten. 1979.

Vol. 751: Number Theory, Carbondale 1979. Proceedings. Edited by M. B. Nathanson. V, 342 pages. 1979.

Vol. 752: M. Barr, *-Autonomous Categories. VI, 140 pages. 1979.

Vol. 753: Applications of Sheaves. Proceedings, 1977. Edited by M. Fourman, C. Mulvey and D. Scott. XIV, 779 pages. 1979.

Vol. 754: O. A. Laudal, Formal Moduli of Algebraic Structures. III, 161 pages. 1979.

Vol. 755: Global Analysis. Proceedings, 1978. Edited by M. Grmela and J. E. Marsden. VII, 377 pages. 1979.

Vol. 756: H. O. Cordes, Elliptic Pseudo-Differential Operators - An Abstract Theory. IX, 331 pages. 1979.

Vol. 757: Smoothing Techniques for Curve Estimation. Proceedings, 1979. Edited by Th. Gasser and M. Rosenblatt. V, 245 pages. 1979.

Vol. 758: C. Năstăsescu and F. Van Oystaeyen; Graded and Filtered Rings and Modules. X, 148 pages. 1979.